HINDUISM

Foundations through
Logic, Math, and Science

ABOUT THE COVER

The symbol on the front cover features an OM within a circle with a dot. OM represents consciousness – *Atman/Brahman*. The circling dot evokes the image of an electron orbiting a nucleus, symbolizing matter.

This symbol invites multiple interpretations, including:

- **Consciousness precedes matter** – as proposed by *Advaita Vedanta*, where *Brahman* is the substratum from which the material world unfolds.
- **Matter precedes consciousness** – as suggested by modern science, where consciousness emerges from complex arrangements of matter.
- **An atom of *Brahma*** – In Hindu mythology, *Brahma* (distinct from *Brahman*) is the deity of creation. The symbol may be seen as an atom of *Brahma: Brahma + Anu = Brahmaanu*, where *Anu* means atom in Sanskrit.
- **Panpsychism** – The symbol also reflects the view that consciousness is an intrinsic and fundamental aspect of all matter.
- **Cycle of causality (*samsara*)** – The circle and dot represent the endless loop of material causation in which humans are entangled as reincarnated beings. The spiritual goal is to transcend this cycle and realize the causeless cause – *Brahman*.
- **Divinity within** – The symbol affirms the presence of the divine in each of us, echoing the profound statement of Hinduism: *Tat Tvam Asi* – "That Thou Art."

In the context of Hinduism, what does the symbol mean to you?

HINDUISM

Foundations through
Logic, Math, and Science

Anand Joglekar

Hinduism: Foundations through Logic, Math, and Science
Copyright © 2025 Anand Joglekar

All rights reserved. No part of this publication may be reproduced, stored in a retrieval system, or transmitted in any form or by any means, electronic, mechanical, photocopying, recording, or otherwise, without the prior written permission of the author, except for brief quotations used in reviews, articles, or scholarly works.

This book is a work of non-fiction. While every effort has been made to ensure the accuracy and reliability of the information presented, the author and publisher make no representations or warranties with respect to the completeness or accuracy of the contents of this book and specifically disclaim any implied warranties. The author and publisher shall not be liable for any loss of profit or any other commercial damages, including but not limited to special, incidental, consequential, or other damages.

The views and opinions expressed in this book are those of the author and do not necessarily reflect the official policy or position of any organization, institution, or other individual.

First Printing: December 2025
First Edition

Paperback ISBN: 978-1-955541-71-8
eBook ISBN: 978-1-955541-73-2
Hardcover ISBN: 978-1-955541-72-5

Library of Congress Control Number: 202592171

Published by **FuzionPress**
1250 E 115th Street
Burnsville, MN 55337
612-781-2815
www.fuzionpress.com

For our grandchildren
Tej, Raina, and Surya

TABLE OF CONTENTS

	Preface	9
	Acknowledgements	11
	Introduction	13
1	Searching for Bliss and Meaning	23
2	Brahman, Atman, and Tat Tvam Asi	39
3	Cause and Effect	65
4	The Creation Equation	83
5	Structure of a Human Being	107
6	Theory of Reincarnation	131
7	Karma Philosophy	151
8	Mathematics of Chatur Varna	167
9	Sthitaprajna and Robust Design	185
10	Purposeful and Happy Life	205
	Glossary	223

PREFACE

Have you ever wondered – *Where did everything come from? Who am I? Why does life unfold the way it does? What is the purpose of life? What happens when we die?* If so, you are not alone.

These are the kinds of questions that drew me to Hindu philosophy – but my journey was not quick or easy. I studied the *Bhagavad Gita* one stanza at a time, guided by teachers whose insights gradually unveiled deeper meaning. Still, I longed for something more structured – a clear, logical map of the terrain. I wished for a book that explained the foundations of Hinduism using the tools we use to learn today – logic, mathematics, and science. A concise book one could read quickly and emerge with clarity, confidence, and curiosity sparked anew.

This is that book.

It is written for thoughtful explorers from all walks of life – anyone curious about the interface between ancient wisdom and modern analysis. Whether you're a beginner or a lifelong student, this book will help you engage with Hindu philosophy in a way that is insightful, practical, and deeply meaningful.

Its structure is inspired by Euclid.

We begin with a few core assumptions and derive their consequences logically and progressively. From *Brahman and Atman to creation, karma, reincarnation, and the pursuit of a purposeful and happy life,* every concept is unpacked step by step, grounded in reason, and illustrated through examples drawn from mathematics, science, and everyday life – a sincere attempt to reconstruct these timeless ideas using contemporary thinking.

The result offers depth without sacrificing clarity or approachability. Each chapter stands on its own while also building toward a complete picture. You will find frameworks for understanding Hindu philosophy more deeply, practical tools for personal reflection and improvement, and perhaps a new lens through which to see reality itself.

I hope this book invites you not only to understand Hinduism's foundations, but to experience the quiet transformation that can come from seeing them clearly. May it serve as a steppingstone to a deeper understanding of this timeless philosophy.

Anand Joglekar
Keller, Texas, USA

ACKNOWLEDGMENTS

A well-known Sanskrit *shloka* (stanza) describes the four ways one acquires knowledge:

आचार्यात् पादमादत्ते पादं शिष्यः स्वमेधया। पादं सब्रह्मचारिभ्यः पादं कालक्रमेण च॥

One-fourth from the teacher, one-fourth from one's own intelligence, one-fourth from fellow students, and one-fourth through the passage of time.

This ancient wisdom perfectly captures how my own understanding of Hinduism has evolved – through a rich tapestry of guidance, self-reflection, peer dialogue, and lived experience.

I am deeply grateful to the teachers who have illuminated this path. The writings and teachings of individuals such as Swami Vivekananda, Swami Chinmayananda, Swami Parthasarathy, and Swami Sarvapriyananda have offered unique insights into Hindu philosophy. Their ability to make ancient wisdom accessible to modern seekers has been transformative.

My heartfelt thanks go to Dr. Shashikant Sane and Professor Prasanna Kumar, whose years of dedicated discourses on the *Bhagavad Gita* at the Hindu Temple of Minnesota first kindled my passion for deeper study. I extend my gratitude to the entire temple community for creating a space where such learning could flourish.

I am grateful to the many readers who generously offered their time to review the manuscript or portions of it. Their thoughtful feedback, suggestions, and encouragement have strengthened this work. I also acknowledge the assistance of artificial intelligence tools, which proved invaluable in helping me refine my arguments.

I am profoundly grateful to my wife, Jaya Joglekar, for many enriching discussions and for her unwavering encouragement from inception to completion of this project.

My silent but deep appreciation extends to all others – authors, speakers, fellow seekers, and even people in casual conversations – who have unknowingly contributed to my understanding of the subject. In the spirit of the ancient *shloka* above, this work is not merely an individual effort, but represents my understanding of the wisdom of all who made it possible.

INTRODUCTION

For millennia, Hindu philosophy has guided those seeking truth, purpose, and happiness. It offers profound insights into consciousness, causality, and the ultimate nature of reality. Yet for today's readers – especially those shaped by logic, math, and science – its traditional presentation can feel elusive, and difficult to grasp. This book aims to change that.

A Logical Approach

This book explores foundations of Hindu philosophy through a logical, mathematical, and scientific lens. Rather than begin with complex scriptural verses that require extensive interpretation, we start with clearly stated assumptions and examine their logical consequences.

Traditional study of Hindu philosophy often begins with metaphorical passages like "Weapons cannot cut It, fire cannot burn It" from the *Bhagavad Gita,* leaving students to derive fundamental concepts through interpretation. This approach, while rich and profound, can be challenging for those seeking systematic understanding. It resembles asking a student to derive Newton's law of gravity from observing an apple and a leaf fall together, without first establishing the conceptual framework of mass, force, and acceleration. Just as we need physicists like Newton to formulate precise laws from natural observations, we need philosophers of Adi Shankaracharya's caliber to systematize the insights embedded in scriptural metaphors.

We take a different approach, similar to how modern geometry textbooks present Euclidean principles. Instead of starting with complex proofs, we state basic assumptions clearly and explore what follows from them. This method makes the logical structure of Hindu philosophy more visible and accessible.

Five Key Assumptions

We examine Hindu philosophy through the following five key assumptions:

1. *Brahman*: Ultimate reality characterized by existence, consciousness, and infinity
2. **Causality**: Every effect has a cause, and every cause produces effects
3. *Prakriti*: Primordial matter with three fundamental qualities (*gunas*) that govern manifestation
4. **Human Structure**: Multiple levels of being, from physical body to *Atman*
5. **Reincarnation**: The soul's transmigration through different lifetimes

By accepting these assumptions and following their logical implications, we can understand foundational concepts of Hinduism – such as *Brahman* (ultimate reality), *Atman* (Self), *Tat Tvam Asi* (That Thou Art), creation, human composition, rebirth, *karma* (action), *varna* (classification), *Sthitaprajna* (person of steady wisdom), and *Karma Yoga* (selfless action) – and how they interrelate to form a cohesive philosophical system.

What Makes This Book Unique

Clarity: Complex concepts emerge through logical steps rather than interpretive leaps, making the reasoning transparent. It helps demystify Hinduism.

Accessibility: No prerequisite knowledge of Sanskrit or traditional commentary is necessary. A comprehensive glossary explains all Sanskrit terms used.

Systematic understanding: Readers can see how different concepts relate to each other and form a complete, consistent system.

STEM perspective: The book uses mathematical and scientific thinking throughout, with mathematical modeling used to illuminate concepts like creation, *Karma Yoga*, *Varna*, and *Sthitaprajna*.

Real-Life relevance: Each chapter explains how these ideas connect to real life and how they can be meaningfully applied.

The book is intentionally concise, designed as an introduction to Hinduism's philosophical foundations. It prepares readers for deeper study of primary texts like the *Upanishads*, *Bhagavad Gita*, and *Brahma Sutras* by providing the conceptual scaffolding needed to appreciate their subtle arguments.

This logical approach complements rather than replaces traditional study methods. Just as learning basic mathematical principles helps students appreciate advanced proofs, understanding the basic foundations of Hinduism can enhance appreciation for its depth and subtleties.

We focus primarily on *Advaita* (non-dual) *Vedanta*, and on *Sankhya Philosophy* (dualism), drawing from the *Bhagavad Gita* and *Upanishads*. Our goal is to provide a systematic way for contemporary readers to begin to understand these teachings.

What You Will Discover
Through ten chapters, this book examines:

- How *Brahman* (universal consciousness) and *Atman* (individual identity) interrelate and what the famous declaration *Tat Tvam Asi* (That Thou Art) means
- How our everyday experiences with cause-and-effect illuminate the foundations of Hinduism, including *karma*, rebirth, and ethical responsibility

- How math and science help clarify metaphysical ideas like creation and *Karma Yoga*
- How philosophical insights from *Gita* and *Upanishads* can be applied to modern life
- How timeless concepts connect with contemporary scientific theories
- How this entire framework becomes livable, not just learnable

Whether you are a curious beginner or a seasoned seeker, this book offers a gateway into Hindu philosophy that is clear, systematic, and powerfully relevant. It honors the tradition's depth while inviting fresh engagement – from minds drawn not just to mystery, but to meaning.

Hinduism: An Overview

Hinduism is one of the world's oldest, and philosophically sophisticated religious traditions. It has over one billion followers across the globe. It is known in Sanskrit as *Sanatana Dharma* – the eternal way of life – doing what is right, living spiritually, and staying connected to the universe. Hinduism differs from many religions in having no single founder, central authority, or uniform creed. Instead, it represents a rich tapestry of beliefs, practices, and philosophical insights that have evolved over thousands of years.

Origins and Unity in Diversity
Hindu tradition views its core teachings not as human inventions but as eternal principles discovered by ancient sages (*rishis*) through direct spiritual insight and passed down through carefully maintained oral and written traditions. This perspective helps explain Hinduism's remarkable diversity: if the underlying truths are universal and eternal, they can be expressed through countless cultural forms while maintaining their essential validity.

Across India's vast landscape, Hindu practices vary dramatically – different languages, customs, deities, and rituals flourish in different regions. One devotee in India may worship Lord Ganesh, another Murugan, while a third might emphasize Shakti worship. Yet beneath this diversity lies a shared philosophical foundation that has provided cultural coherence for millennia.

These underlying principles have proven remarkably resilient, surviving invasions, foreign rule, and systematic attempts at suppression. They continue to shape not only religious practice but also India's art, literature, social structures, legal concepts, and cultural identity. Understanding these foundational principles forms the primary subject matter of this book.

Pathways to Fulfillment
Hinduism offers multiple systematic approaches to spiritual growth and life fulfillment, recognizing that different personalities require different ways. The tradition identifies four primary paths, each designed to work with specific human temperaments.

Karma Yoga emphasizes selfless action – performing one's duties and responsibilities without attachment to outcomes. This path appeals to those who find meaning through service, work, and engagement with the world.

Bhakti Yoga focuses on devotion, and surrender to the divine through worship, prayer, ritual, and emotional connection. This path suits those drawn to love and devotion. It recognizes that the heart's capacity for love can be a powerful vehicle for spiritual transformation.

Jnana Yoga pursues wisdom through study, contemplation, and intellectual inquiry into the nature of reality. This appeals to those with philosophical inclinations who seek to understand the logical structure underlying existence. It emphasizes discrimination between the eternal and the temporal.

Raja Yoga develops mental discipline through systematic meditation and yogic practices, attracting those who seek direct experiential knowledge. It provides practical techniques for transcending ordinary mental limitations.

These paths are not mutually exclusive; most practitioners naturally combine elements from different approaches according to their temperament, life circumstances, and stage of development. The tradition recognizes this integration as both natural and beneficial.

Foundational Texts

Hindu philosophy draws from extensive literature spanning thousands of years, with three categories of texts forming its authoritative core:

The *Upanishads* are the concluding section of the *Vedas*, Hinduism's oldest and most authoritative scriptures. These philosophical treatises explore fundamental questions about reality, consciousness, and the non-duality of *Atman* and *Brahman*. The word "*Upanishad*" means "sitting near," reflecting the intimate teacher-student relationship through which this wisdom was traditionally transmitted. Many *Upanishads* exist, the traditional number often cited being 108. About a dozen are considered principal texts that form the foundation of *Vedantic* philosophy.

The *Bhagavad Gita*, or "Song of the Divine," presents Krishna's teachings to the warrior prince Arjuna on the eve of the Kurukshetra battle. Through this dialogue between the divine teacher and the sincere student, the text addresses practical questions about duty, action, and spiritual development that remain relevant for contemporary readers. It masterfully integrates the theoretical insights of the *Upanishads* with practical guidance for ethical living in complex social situations.

The *Brahma Sutras* systematize the *Upanishads'* diverse teachings into a coherent philosophical framework, addressing potential objections and clarifying concepts that might otherwise seem contradictory across different texts. Attributed to Badarayana (also called Vyasa), these concise aphorisms require extensive commentary for full understanding, leading to the rich tradition of *Vedantic* interpretation.

Together, these texts form the three pillars of *Vedantic* philosophy, each serving a distinct but complementary function in philosophical education.

Philosophical Schools

Hinduism encompasses six classical schools of philosophy, all accepting the *Vedas* as authoritative while emphasizing different aspects of truth and practice. These schools – *Sankhya, Yoga, Nyaya, Vaisheshika, Mimamsa, and Vedanta* – represent sophisticated philosophical systems that offer different interpretations of the underlying truth.

Sankhya, attributed to Sage Kapila, is the oldest of these schools and is a dualistic system of Hindu philosophy that posits infinite eternal independent realities: *Purushas* (infinite pure consciousnesses) and *Prakriti* (primordial matter). Liberation (*moksha*) is achieved when *Purusha* realizes its distinction from *Prakriti*, ending the cycle of suffering and rebirth.

The most influential today is *Vedanta*, which focuses on the ultimate nature of reality and consciousness as revealed in the *Upanishads*. Within *Vedanta*, three major interpretations have emerged:

Advaita (non-dualism), primarily systematized by Adi Shankaracharya (8th century CE), teaches that individual consciousness (*Atman*) and universal consciousness (*Brahman*) are identical. Apparent differences result from ignorance that can be removed through knowledge and practice.

Vishishta Advaita (qualified non-dualism), formulated by Ramanujacharya (11th – 12th century CE), proposes that individual souls remain distinct yet depend entirely on *Brahman*, like waves depend on the ocean while maintaining their apparent individuality.

Dvaita (dualism), developed by Madhvacharya (13th century CE), maintains the eternal distinction between individual souls and the Supreme Being, emphasizing devotional relationship rather than identity.

Each school provides internally consistent answers to fundamental philosophical questions while acknowledging the validity of different approaches for different seekers.

This book primarily deals with non-dual *Advaita Vedanta* with *Brahman* as the single ultimate reality. Where necessary, a dualistic *Sankhya Philosophy* is considered with *Brahman* and *Prakriti* as two independent ultimate realities. In this formulation, the infinite *Purushas* are considered as a single infinite *Brahman*.

Contemporary Relevance

Hindu philosophy continues to offer valuable insights for modern life and thought. Its core teaching invites us to ask, "Who am I, really?" – beyond roles, emotions, and accomplishments. In an age of identity confusion and anxiety, this introspection is more than philosophical – it is grounding. Hinduism reframes "why things happen" not as punishment or reward, but because of systematic unfolding of causality and as an evolutionary process aimed at achieving liberation. This supports ethical behavior, growth mindset, and systems thinking – essential in both personal and societal life.

Its core teaching, *Tat Tvam Asi* – That Thou Art – affirms that divinity is within each of us. When this truth is lived, one realizes that the whole world is our family. In a time of fragmentation and division, this is not just a spiritual insight – it is a healing force.

Its emphasis on multiple valid paths (religious pluralism) respects individual psychological differences while maintaining shared ethical values. The multiple paths provide a personalized "operating system" to lead a purposeful and happy life.

Hinduism's sophisticated analyses anticipate contemporary discussions in science and philosophy. Both investigate the origins of the universe – whether it is a beginningless evolving universe and whether time flows linearly. Both wrestle with whether consciousness is fundamental or emergent – and how it shapes reality. They agree that reality is not what meets the eye, it is deeper, subtler, and often hidden. Both recognize that effects often emerge from complex, interdependent causes – not always predictably. Both affirm that humans can evolve – mentally, ethically, and spiritually—through disciplined effort.

Most significantly, Hinduism's core insight – that diverse approaches can lead to the same fundamental truths – offers a practical model for navigating our increasingly interconnected yet culturally diverse world. In recognizing both unity and diversity as essential aspects of reality rather than contradictory opposites, Hindu philosophy provides a mature framework for understanding how different perspectives can coexist, enrich each other, and contribute to collective human flourishing.

This ancient tradition's ability to adapt while maintaining its essential character demonstrates the enduring relevance of its foundational insights for seekers of meaning in any era.

CHAPTER 1

SEARCHING FOR BLISS AND MEANING

A nand – whose name means "bliss," though he admits it still eludes him – takes readers on an unexpected journey that begins with a simple choice: to go to the local temple. What unfolds is a story where Sanskrit verses become equations, varying translations uncover deeper truths, and a pothole-riddled road with a Dead-End sign becomes a metaphor for interpreting one of the world's oldest traditions.

This is a thoughtful exploration of Hinduism for minds shaped by reason, yet still drawn to the sacred. If you have ever found yourself standing at the crossroads of tradition and modernity, this essay invites you in – with insight, humility, and a map for the journey ahead.

The essay was written years ago and describes the authors initial thinking and experiences while starting to study Hindu philosophy. Readers may have had similar experiences. The temple referred to in this essay is the Hindu Temple of Minnesota, located in Maple Grove, a suburb of Minneapolis, Minnesota, USA, as it was then.

I finally decided to go to the temple.

My name is Anand. It is a common name in India. It means bliss. When asked as to why I go to the temple, I often reply by saying that I am in search of two long-lost friends, named *Sat* and *Cit* (pronounced chit). *Sat* means absolute existence. *Cit* means absolute consciousness. The Hindu God – ultimate reality – is known as *Brahman* and Hindus believe that *Brahman* is *SatCitAnanda* – existence, consciousness, bliss. I have yet to find my two friends, *Sat,* and *Cit*. Not only that, but I am also far from having found bliss. Hindu philosophy says that *SatCitAnanda* is within me. I must continue to search inward.

I do not go to the temple because God resides only there. A devotional singer once told this story. A beggar went begging at the temple. The temple was open from 6 AM to Noon, and then from 4 PM to 8 PM. He collected twenty rupees for his effort. On the way back to his hut, he decided to stop by the local bar. Between 9 PM and midnight, he begged outside the bar and collected two hundred rupees. On the way home, he thought to himself, "God, what a surprise? You give me one address but live somewhere else!"

This essay explores my journey to understand Hindu philosophy through three distinct but interconnected experiences: first, the physical and symbolic journey to the temple itself; second, the translation and interpretive challenges I encountered when studying sacred texts like the *Bhagavad Gita*; and third, my discovery of how mathematical and analytical thinking can illuminate ancient wisdom. Together, these experiences illustrate the approach I have adopted – bridging traditional understanding with modern analytical tools while respecting the depth and complexity of Hindu philosophy.

Journey to the Temple

This first experience – the physical journey itself – became a metaphor for the interpretive challenges that lay ahead.

Our temple in Maple Grove, a suburb of Minneapolis, Minnesota, USA is 20 minutes away from my home by car. As I take Troy Lane to go to the Hindu Temple of Minnesota, I notice that the road is long but straight. I can see the temple to the right in front of me. The road itself, while straight, is not smooth. It has many potholes, rough spots, and ups and downs. Is the road worn because of the large number of cars going to the temple – the thousands of devotees that come to the temple? Was it left in disrepair because of neglect or lack of resources? Or was it left that way to suggest that while the path to God is straight, it is full of man-made obstacles?

When Troy Lane comes to an end, I see the temple parking lot to the right. There is a shortcut that I could take to get to the parking lot. There is no sign to tell me that this is not what I am supposed to do. I know that instead of taking the shortcut, the idea is to turn right, and then left, and left again, and then complete a half circle clockwise around the temple to the parking lot. While leaving, one is to continue going clockwise and thus complete a full circle around the temple. Completing such circles clockwise around the temple is considered auspicious. It is a tradition in India.

Most devotees are in a rush and invariably take the shortcut to and from the parking lot. Clearly marked signs would help. But providing such clear dictates and insisting upon only one path to follow has not been the Hindu tradition.

While driving to the temple, when I enter Troy Lane, I see a sign on the road that says, "Dead End." Right under this sign is the sign for the Hindu Temple. What does this mean? Does it signify that the Hindu Temple is a dead end, or does it mean that when people reach a dead end in their life and, having tried everything else, they turn to the Hindu Temple? Was it merely a matter of convenience to hang the Hindu Temple sign on a pre-existing Dead-End sign? Or does it mean something entirely different?

Interpreting *Bhagavad Gita* – the Divine Song – a cornerstone of Hindu philosophy, seems like answering such questions. The interpretation is in the eyes of the beholder.

Most devotees visit the temple and head upstairs, where there are mini temples dedicated to many Gods and Goddesses – Vishnu, Shiva,

Ganapati, Sarasvati, Ram, Krishna, and others. They pray to their chosen deity – often seeking a boon. This reminds me of a story.

A man performed penance for a long time, and God appeared before him. The man asked God, "How long is a million years in your terms?" God replied, "One minute." The man then asked, "How much is a billion dollars in your terms?" God replied, "One dollar." The man said, "Please give me one dollar," and God said "Okay, but wait a minute."

Discourses

My purpose for going to the temple is to attend discourses on the *Bhagavad Gita* and Hindu philosophy, held on Sunday mornings from 10:30 AM to Noon in the auditorium. Lunch follows in the cafeteria across from the auditorium. What distinguishes these discourses from similar events is the willingness of the speakers to answer audience questions. One of the speakers insists on questions, even calling on members of the audience by name. For me learning a subject without questions, answers, and discussion is challenging. Often the conversation continues over lunch.

What really draws me there? Is it the discourses, the questions and answers, or the lunch? These three aspects cater to the *gunas* inherent in us. *guna* means quality. There are three types of *gunas*: *sattva,* meaning pure and serene; *rajas,* meaning passionate and agitated; and *tamas,* meaning dull and inactive. These tendencies shape our behaviors and experiences. The discourses, questions and answers, and lunch cater to these qualities.

Questions and answers, along with the more detailed discussions that sometime occur over lunch, are unusual features of these sessions. The questions people ask and the answers that follow are truly illuminating. I am continually amazed at the ability of speakers to cogently answer these questions. This is not an easy task and requires a clear understanding of the fundamentals. I wish there were even more time for questions and answers and that people asked more questions. Perhaps others understand it all, or think that they can figure it out on their own. Many shy away from asking questions because they fear that they will look foolish.

It is not that I am following the dictates of the *Bhagavad Gita* in my personal life – do as I say and not as I do, I always say. But this does not prevent me from asking questions. I am grateful that the speakers are indulgent in this matter. At least, that is what I thought until one of the speakers told me that I must provide at least one answer to each question I ask.

He was asking – when will the questions become a quest? That is partly why I started writing these essays.

The Challenges

As I started learning Hinduism, I encountered a second experience: making sense of translation from the original Sanskrit, and the varied interpretations offered by writers and speakers. The interpretive journey was not unlike the physical journey to the temple itself.

Learning Hinduism
Religion requires faith. Our dog Rocky has faith.

Faith is the dog, who sees another,
And barks hopefully, sight unseen.
~Based on Tagore's "bird"

Most religions have their God, their metaphysics, and their principles for a happy and purposeful life. Religions often disagree regarding God and metaphysics – the notion of "my God is better than your God" has led to countless conflicts and even millions of deaths. It is the principles for purposeful and happy life that seem the most useful, if implemented. *Then why have God and metaphysics?*

Some believe that God and metaphysics, as stated in the scriptures, are the absolute truth. Others believe that the concepts of God and metaphysics, formulated thousands of years ago, were the best available explanations for the world around us but are now myths. Some think that God and metaphysics were invented to motivate moral behavior in the absence of the rule of law. Still others believe that God and metaphysics are a deliberate fabrication created by humans to control the masses for the benefit of the upper echelons of society. Some say that God represents the sum of human ignorance at any given time, as the phrase "God of the gaps" suggests.

My purpose in going to the temple is not to debate these judgments. My goal is to attempt to understand Hindu God, metaphysics, and principles for a happy and purposeful life.

I am not a theologian and do not believe that God dictated Hindu metaphysics. I believe that Hindu metaphysics is based upon the profound thinking of great Hindu philosophers. The interpretation of this philosophy has, knowingly and unknowingly, shaped the thinking of more than a billion Hindus and continues to do so. Hindu philosophy has been one important factor that has kept a diverse country together for millennia – diverse in terms of languages, food, clothing, appearance, culture, and many other factors. It is therefore important to understand Hindu metaphysics.

The fact that these metaphysical ideas have survived for several thousand years means that they are capable of being reinterpreted in the context of changes that have taken place, while providing useful practical guidance. Such great philosophers as Adi Shankaracharya, Ramanujacharya, and Madhvacharya have offered varying interpretations of such fundamental ideas as the Hindu God *Brahman* and the non-duality of *Brahman* and *Atman*. *Atman* is the *Brahman* within us. Interpreting scriptures is not unlike trying to divine the original intent of the writers of the United States constitution, with many five-four decisions in the Supreme Court.

As I started attending discourses, I realized that I know little about Hindu philosophy, with a superficial understanding of the *Bhagavad Gita* and no understanding of the *Upanishads* – the philosophical texts of Hinduism. I was not even aware of the existence of *Brahma Sutras* – aphoristic verses dealing with *Brahman*. I could not spell Schopenhauer without a spell checker. In other words, I was like the many others who, though ill-equipped, wish to understand Hinduism. How to proceed?

It seems to me that one must start with *some* understanding gained by listening to people who know the subject – in person or otherwise, reading relevant literature, and thinking about what one has learned. However, the material can be complex – originally written in Sanskrit in poetic form. Sometimes, the translations and illustrative examples people provide, though well intended, can leave a wrong impression that can be hard to erase. Initially, it seems best to listen carefully, think hard, but not let what is being said to be unduly etched too severely on an open, receptive, and blank mind.

After *shravana* – listening, and *manana* – thinking about what one has heard and read, the next important step is to *write down* what one has understood so far. The process of writing down, thinking about what one has written, discussing it with others, and looking for factual and logical inconsistencies is a good way to continue learning. Clearly stating one's ignorance of a subject sharpens and accelerates learning; and leads one to properly value what others say. It is then possible to evaluate additional information to see where one has gone wrong, make the necessary adjustments, and continue to make progress.

Krishna advices as such in the *Bhagavad Gita,* Chapter 18, Verse 63:

> *You have heard*
> *The secret knowledge*
> *Reflect on it, and then*
> *Act as you will*

He says, "Think about what I have said and then do what you think is right." Great teachers would be disappointed if students follow what the teachers said merely because they said it. Great teachers want to be heard, want their words to be taken seriously, but also want students to think independently, ask questions aimed at clarifying the concepts, and then come to their own conclusions. Otherwise, all progress will stop.

That is why several years ago, I started reading books, and listening to discourses – to attempt to understand Hindu philosophy, to think about what I have heard, to write what I have understood, and to discuss with others what I have written to improve my understanding. Over the last few thousand years, it is unlikely that major evolutionary changes have occurred in human beings. The main changes are the exponential growth of science, technology, engineering, and mathematics (STEM) and their effects on human life. STEM is the new *pramana* – means of knowledge. Any interpretation of scriptures needs to consider these changes so that Hindu philosophy can be better communicated to the audience of today, in today's language, and context. That is my interest in writing essays.

Hindu philosophers seem to say that such basic concepts as *Brahman* and *Atman* cannot be sensed, described, and understood by our sensory organs, mind, and intellect. For every attempt at description in an understandable fashion, *neti, neti, neti* – not this, not this, not this – is the answer. This may well be, but it can be very frustrating.

As one practical example, let us suppose that you go to the doctor complaining of a stomachache. You ask the doctor whether it might be due to something you ate the previous night. The doctor says *neti – not that*. You ask him whether you may have stomach ulcer. Again, he says *neti – not that*. You ask him whether you may have stomach cancer. Again, he says *neti – not that*. For every medical problem that you encounter, if *neti* is always the answer, and he never comes around to saying *iti, iti, iti – this, this, this;* you are likely to find another doctor.

One teacher at the temple once said that he interprets *neti* as *"not only that."* I thought to myself that this certainly makes matters worse. If I ask the doctor whether I have stomach ulcer, it will be worse if the doctor says, *"not only that."*

It took me time to understand that defining *neti* as *not that* is to point out that my body, mind, and intellect are not *Brahman*. It points out that what I usually think of as "me," namely my body-mind-intellect (BMI) is different from *Brahman*. This sets up the duality of *Brahman* and our body-mind-intellect, leading to the *dvaita* (dual) philosophy of Hinduism.

On the other hand, defining *neti* as *"not only that"* means that everything is *Brahman*. This leads to the *Advaita* (non-dual) philosophy of Hinduism.

The above analogy is not perfect. Medical diagnosis requires specific identification, while Hindu philosophy's "neti, neti" serves a different purpose of pointing to the ineffable nature of ultimate reality. Even so, it is frustrating to realize that the overall message is that *Brahman* cannot be understood by our senses, mind, and intellect. If so, then one should quit listening, thinking, and talking about God– if the purpose is to understand God. Entirely different means of understanding would be necessary. This would be bad news for me and for many others, because as Descartes said – "I think, therefore I am." Or, as *Advaita Vedanta* would have it – "I am, therefore I think!" One suggests the supremacy of the mind, and the other suggests the supremacy of *Atman*.

Fortunately, some basic tenets of Hindu philosophy have been described, thereby opening the door to further understanding. It could be, and has been, argued that God and metaphysics are in the eyes of the beholder. This is not necessarily completely accurate because every conceptualization does not conform to the basic tenets of Hindu philosophy. For example, statements like – "God created the universe by magic. He can turn gravity downside up. There is no telling what will happen next." Or "God created the universe in six days and rested on the seventh" – do not capture Hindu philosophy.

In discussing such matters, it is important to recognize that all languages are imperfect in communicating the exact meaning of words. Theologians may think that every word in *Bhagavad Gita* is the word of God, but they do not know what every word in *Bhagavad Gita* really means. Words are our friends and our enemies. That is why mathematics is the preferred language of physics. Why this physical world should be understandable in mathematical terms is an interesting mystery. Is mathematics God?

Bhagavad Gita

Most of the discourses at the temple dealt with *Bhagavad Gita*, the song of God, the famous book of Hindu philosophy that explains the way to purposeful, happy life. The Gita was taught by Krishna, the God incarnate, to Arjuna, the representative of the common person, on the battlefield at Kurukshetra. When the train from Delhi in northern India stops at Kurukshetra, people get off at the station and walk the fields barefoot. It is thought that every human being, in his or her lifetime, breaths at least a molecule of nitrogen that had once passed through the body of Jesus Christ. Some now want to breathe the air and touch the soil that Krishna had once breathed and touched.

Gita was originally written in Sanskrit, but few understand the language today. It has been translated into English, with extensive commentaries, by many learned individuals from the East and the West. Translating from Sanskrit to English is not an easy task. It is an interesting exercise to take an important verse from Gita and look at its many English translations. In some cases, every translation imparts a different meaning when read in English.

One of the most famous and commonly quoted verses in *Bhagavad Gita*, verse 2.47, begins with *"Karmani Eva Adhikarh Te."* Here is how some very well-known and knowledgeable people have translated these four words.

- You have the right to work – Eknath Easwaran
- Your duty is but to act – C. Rajgopalachari
- Thy right is to work only – Chinmayanand
- Action alone is thy province – Mahadeo Desai
- Be intent on action – Barbara Stoler Miller

Each translation imparts a slightly different meaning and one begins to wonder if any of them capture the true meaning.

Given that these translators are all recognized Sanskrit scholars, how does one choose among their interpretations? The key is to consider which translation best aligns with the Gita's central themes. Throughout the text, Krishna emphasizes the importance of controlling one's senses, mind, and desires. The Gita consistently teaches that while we cannot control external outcomes, we can – and must – control our internal responses and voluntary actions.

From this thematic perspective, 'control over actions' emerges as the most coherent interpretation. We indeed have control over our voluntary actions – this is the essence of free will and moral responsibility that makes Krishna's entire teaching relevant. The other interpretations, while linguistically valid, do not capture this crucial element of self-mastery that runs throughout the Gita.

To further understand the translations from a practical perspective I looked up a Sanskrit-to-English dictionary. Sometimes there are half a dozen to a dozen English synonyms for a Sanskrit word, and depending upon the synonym you choose, the meaning imparted by the translation can materially change. *Karma* can be translated as action and as work. *Adhikarh* is usually translated as right but also as control. That makes for four possibilities: right to work, right to act, control over work, and control over actions. Upon reflection it becomes clear that, from a practical perspective (a) "right to work" is invalid because employment is not guaranteed, (b) "right to act" is invalid because our actions are constrained by laws, ethics, and social norms, and (c) "control over work" is invalid because external factors (boss, circumstances) often dictate our work. Clearly the last of the four possibilities is the only practically correct one, and "one has control over one's voluntary actions" becomes the correct choice from various perspectives.

Verse 2.47 correctly translated is:

> *You have control over your actions,*
> *but not on their fruits.*
> *Act for action's sake,*
> *And do not be attached to inaction.*

Mahatma Gandhi encapsulates the central message of Gita in two words: *nishkama karma,* selfless action, work free from selfish desires. Desire is the fuel of life; without desire nothing can be achieved.

Nishkama is selfless desire. *Karma* means action. Gita counsels – work hard in the world without any selfish attachment to the results.

Mahatma Gandhi goes on to say "By detachment I mean that you must not worry whether the desired result follows from your action or not, so long as your motive is pure, your means correct. It means that things will come right in the end if you take care of the means. But renunciation of fruit in no way means indifference to results. Regarding every action, one must know the result that is expected to follow, the means thereto, and the capacity for it. He who being so equipped, is without selfish desire for the result and is yet wholly engrossed in the due fulfillment of the task before him, is said to have renounced the fruits of his action. *Only a person who is utterly detached and utterly dedicated is free to enjoy life. Renounce and enjoy!"* (Adapted from *Bhagavad Gita* by Ekanath Easwaran).

What this means for us is that we should carefully set our goal, and then be utterly focused on the actions necessary to achieve the goal without constantly worrying about whether the goal is being achieved or not. Periodically there needs to be an evaluation of whether the actions we are taking are appropriate or not, and if not, what actions should be taken. It is also conceivable that the goals that we have set are incorrect and need to be changed.

As we go through life in this manner, many factors – social, economic, emotional, physical, bad habits we may have, even traffic jams while going to work – could affect our mental balance and happiness. We should condition ourselves, you may even say, we should redesign ourselves, to keep evenness of mind in the face of all these disturbances. Such a person of steady wisdom is called *Sthitaprajna*.

Mathematical Thinking

The third experience that shaped my approach was discovering how mathematical and analytical thinking could serve as a bridge between ancient wisdom and modern understanding to explain the foundations of Hinduism.

The essays that follow deal with the core concepts of Hinduism as I have understood them. I have taken an *iti, iti, iti* (this, this, this) approach using mathematical thinking rather than the *neti, neti, neti* (not this, not this, not this, not this) approach. This is a useful but simultaneously dangerous approach. The *iti, iti, iti* approach requires creation of mathematical or conceptual descriptions called models – whether approximate mathematical equations, word descriptions, analogies, conceptual examples, or visual illustrations. It is known that all such models are wrong, but some are useful. We know that Newton's law of gravity is approximate, but it is correct enough to be practically useful.

Euclid

My basic approach is like that of Euclid to his plane geometry. Euclid starts with a few definitions and assumptions. For example, a point is defined to have no dimensions. One could argue that this assumption is wrong because if a point has no dimensions, it does not exist. Even so, the assumption is understandable, and we are to accept his assumptions without question. Then all theorems of plane geometry follow as a logical consequence of these assumptions. I want to take a similar approach by identifying the key assumptions of Hinduism and then, based on these assumptions, show how the foundational concepts follow logically.

Mathematics

To illustrate the role simple mathematics can play to enhance our understanding of the subject, consider a famous Sanskrit peace prayer, stated at the beginning of *Isha Upanishad* perhaps more than three thousand years ago.

Sanskrit: ॐ पूर्णमदः पूर्णमिदम् पूर्णात् पूर्णमुदच्यते ।
पूर्णस्य पूर्णमादाय पूर्णमेवावशिष्यते ॥

Translation: Om, That is complete, this is complete, from completeness comes completeness. If completeness is taken away from completeness, only completeness remains.

The second half of the stanza says that if *Poorna* (complete) is taken away from *Poorna* (complete) what remains is *Poorna* (complete).

The interpretation depends upon one's point of view. Consider a red rose as an example. A Romeo looks at a red rose, plucks it, and gives it to his Juliet. A poet looks at a red rose and writes songs of love. A scientist looks at a red rose and asks – How does the eye see red?

Some may look at the stanza and conclude that there must be a mistake. If you take away *Poorna* from *Poorna*, the remainder should be zero.

A theologian knows that before creation there was only God. Nothing else existed – no devil, not even "nothing." In the above stanza, "That" means God and God is obviously complete (*Poorna*). In the stanza, "this" means God's creation, and it is also complete (*Poorna*). God created His creation out of Himself because there was nothing else to create from. God provided the raw material necessary for creation out of Himself. But the amazing thing was that God did not get used up to any extent during the process of creation, and even after the creation, God continued to remain complete (*Poorna*).

God's process of creation is unlike any used by humans. The manufacturing process used by humans is such that the raw material gets used during the process of creation.

A mathematician looks at the stanza and says that it represents the equation:

Poorna – Poorna = Poorna

Poorna is an unknown, and if we call it X, the equation becomes:

$X - X = X$

There are two values of X that satisfy the above equation: $X = 0$ and $X = \infty$ (infinity). It is easy to see that $0 - 0 = 0$ and mathematically ($\infty - \infty$) is indeterminate, meaning that the answer could be any value, including ∞ itself.

$X = \infty$ means that God is infinite. It then follows that God's creation is also infinite. We will extensively discuss the meaning of $X = 0$ in the chapter "The Creation Equation."

Mathematical thinking clarifies and adds to the theological interpretation of the stanza and supports the conclusion that God is infinite and God's creation is infinite. It shows why God did not get used up during the process of creation and continues to be infinite even after creation. Mathematics adds to the theologian's understanding, rather than subtracting from it. This demonstrates how modern tools can illuminate ancient wisdom.

In each chapter to come, we will take a similar approach of starting with a key assumption of Hinduism, and explaining its meaning and implications using mathematics, science, and logic. Non-standard interpretations will also be discussed where they make sense. The goal is to quickly get a first level understanding of the foundations of Hinduism.

Summary

This opening chapter offers a fresh perspective on dealing with Hindu philosophy, integrating modern analytical thinking with ancient wisdom. The chapter unfolds through three dimensions of learning.

First, the physical journey to the Hindu Temple of Minnesota – with its pothole-ridden road, "Dead End" sign, and unclear directions – becomes a powerful metaphor for the interpretive challenges inherent in understanding Hindu thought. The journey mirrors the intellectual and spiritual path of any sincere seeker, emphasizing the importance of navigating uncertainty with perseverance.

Second, the examination of varying translations of Sanskrit texts – such as *Bhagavad Gita* verse 2.47 – reveals how seemingly minor differences in word choice create dramatically different practical meanings. The essay demonstrates the importance of questioning interpretations systematically rather than accepting them merely because they come from recognized authorities.

Third, applying mathematical reasoning to the *Isha Upanishad's* peace prayer introduces an unconventional yet illuminating approach, showing how the mathematical equation $X - X = X$, where X represents infinity, can enhance theological understanding rather than diminish it.

These learnings lead to a proposed methodology adopting the "*iti, iti, iti*" (this, this, this) approach instead of the traditional "*neti, neti, neti*" (not this, not this, not this) method. The specific approach, taken in subsequent chapters, is to use a Euclid-like method – identify key assumptions of Hindu philosophy, and follow their consequences. This methodology neither dismisses tradition as outdated nor accepts it uncritically, instead proposes the use of logic, mathematics, and science to make timeless concepts accessible to minds shaped by STEM thinking.

CHAPTER 2

BRAHMAN, ATMAN, AND TAT TVAM ASI

"*You do not have to die to meet God. You can realize God while living because God is within you*" – so says Hinduism. Divinity is our very essence.

Hinduism, particularly the Advaita Vedanta philosophy, recognizes one ultimate reality called Brahman defined as Satyam (existence), Jnanam (consciousness), and Anantam (infinite). This is the first fundamental assumption of Hinduism. This essay explains the definition of Brahman in detail, examines its consequences, and discusses its practical implications including the great saying, Tat Tvam Asi – That Thou Art, revealing how the path to spiritual realization lies in the recognition of one's own nature.

The *Brahman* Assumption

Hinduism, particularly the *Advaita Vedanta* (nondual) philosophy, recognizes only one ultimate reality – *Brahman*, as the one and only first cause of the universe.

Brahman is said to be inconceivable, something that cannot be understood or grasped by thinking alone. Therefore, *Brahman* is usually explained by negation – *neti, neti, neti* – not this, not this, not this, meaning that *Brahman* is not like any object we are familiar with. Everything that we can perceive with our senses or think about with our mind, such as thoughts, is an object. *Brahman* is not an object; it is the subject. An object cannot understand a subject. Just as our physical body cannot understand our mind, our mind cannot understand *Brahman*. Trying to understand *Brahman* can be very frustrating.

In this essay, we are attempting to conceive the inconceivable. We will take an approach like that taken by Euclid for his plane geometry. You may remember plane geometry, fondly or otherwise, from your middle school days. Euclid uses definitions, axioms, postulates, and assumptions and based on these, plane geometry logically unfolds one theorem at a time. One such definition is that a "line is length without breadth." We could criticize it and say that if it were so, then the line would not exist. We are not to do that. We can certainly imagine a line with no thickness. We are to accept the definition and examine its consequences.

Is it surprising that Euclid defines a line as breadthless length? Not necessarily. If we assume that a line does have thickness, then we will have to define what that thickness is. Also, in computing the area of a rectangle, there would be an inner and an outer area of the rectangle leading to many complications. Therefore, the definition that a line has no thickness seems reasonable.

Taittiriya Upanishad is one of the top 10 *Upanishads* according to Adi Shankaracharya, the great exponent of *Advaita Vedanta*. It says that **Brahman is Satyam (Existence), Jnanam (Consciousness), Anantam (Infinite)**. This is the first major assumption of Hinduism. An assumption is a belief without evidence or proof. We call this an assumption, and not an empirically testable hypothesis, because, as we shall see later, this assumption cannot be experimentally evaluated as we usually evaluate scientific hypotheses. We will not question the assumption, but will attempt to understand what it means. We will then think about the consequences of the assumption. Finally, we will consider the practical implications of the assumption for daily living.

We could have assumed *Brahman* to be anything we wanted. How to choose the right assumption? The *Brahman* assumption should be such that its implications are not directly falsified by evidence. It should have the capability to explain why *Brahman* is the first cause of this universe and how this universe came about.

For example, let us assume that *Brahman* is a magician and can do anything He wants. Then, we would have to define what magic He can perform. Can He do anything He wants, or are there limitations? Can humans understand His magic? And so on. Finally, the implications of this assumption will have to be understood and compared to our practical experience. For example, if He can turn gravity downside up, what does that mean and is it consistent with our practical experience?

As another example, the *Brahman* assumption could have been that *Brahman* is in heaven and helps only those who believe in Him and constantly praise Him. Once again, we would have to define this assumption more precisely – what and where is heaven – understand the implications of this assumption and, whether the implications conform to our practical experience. For example, is it true that people who believe in *Brahman* and constantly praise Him are better off than those who do not?

Hindu philosophers did not make either of these two assumptions. In assuming *Brahman* to be *Satyam, Jnanam, Anantam* (existence-consciousness-infinite), they appear to have been motivated by a desire to understand and explain how this universe came about. This insatiable desire is beautifully captured by the well-known *Nasadiya Sukta* – the hymn of creation – from *Rigveda,* composed thousands of years ago. Here is a portion of the *Nasadiya Sukta.*

Who really knows? Who in this universe may declare it? Whence was this creation, who knows whence it arose?

He from whom this creation arose, He may uphold it, or He may not. He who supervises this creation, He knows, or He knows not.

On the one hand the hymn expresses the feeling that if it is a beautiful exquisitely crafted Swiss watch, there must be a watchmaker. On the other hand, there is the uncertainty as to whether a watchmaker is necessary, and some doubt as to whether one would ever find the truth.

Is it surprising that the *Taittiriya Upanishad* defines *Brahman* as existence-consciousness-infinite? It is our practical experience that this universe is filled with innumerable physical and mental objects (such as thoughts). Some objects are short lived and others live for a long time. What is common to all these objects is that they exist. Some of these objects, such as human beings, demonstrate consciousness. Finally, the universe is large, perhaps infinitely large, and has existed for a long time. If the universe and the objects in it are to have come from *Brahman*, then *Brahman* would have to be the giver of existence, giver of consciousness, and infinite. This is what *Satyam* (existence), *Jnanam* (consciousness) and *Anantam* (infinite) mean.

Let us now understand *Brahman* is greater details, by defining the words *Satyam, Jnanam,* and *Anantam,* starting with *Anantam.*

Ananta

Anantam or *Ananta* means "without end, without limit, limitless." *Brahman* is infinite. In Sanskrit, two words denote infinity: *Asankhya* and *Ananta*. *Asankhya* means countless, innumerable, countably infinite (1, 2, 3, ……. ∞) used in the context of discrete entities. *Ananta* means unending, eternal, endless in time and space, used in the context of continuous entities. *Ananta* is uncountably infinite, the infinity of real numbers that includes numbers such as $\pi = 3.1416$…….in addition to integers. *Ananta* is a larger infinity than *Asankhya*. That one infinity could be larger than another, was mathematically proved by Georg Cantor a little over a hundred years ago. This is not to suggest that Hindu philosophers knew of Cantor's work, but it is interesting that they, thousands of years ago, used *Ananta* and not *Asankhya* to describe *Brahman*.

Brahman is infinite in three ways: infinite in space, infinite in time, and infinite in objects.

Infinite is space: This means that *Brahman* is infinite and everywhere. Infinite does not mean everywhere. For example, the sequence 1, 2, 3, ……∞ is infinite but it is not everywhere. It does not include the number 1.5. Conversely, everywhere does not imply infinite. That there is air everywhere in this room does not mean that the air in the room is infinite. *Brahman* is both, infinite and everywhere. There is no place where *Brahman* is not.

There are several implications of *Brahman* being infinite and omnipresent. One implication is that *Brahman* cannot move. It has no need to move since It occupies all positions. A second implication is that nothing exists beyond *Brahman*; space itself is within *Brahman*. A third implication is that *Brahman* permeates the entire universe. This is what *Chandogya Upanishad* means by *Sarvam Khalvidam Brahman* – All this is *Brahman*. A fourth implication, as *Bhagavad Gita* points out, is that *Brahman* cannot be cut, burnt, wetted, or dried, namely, cannot be altered. It is not because *Brahman* is made of indestructible material, but because to cut means to create space where *Brahman* is not. This is not possible because *Brahman* is everywhere, always.

Infinite in time: This means eternal, without beginning or end. It means that *Brahman* is unborn and immortal.

Science posits that this universe began about 13.8 billion years ago, and time as we know it, began with the universe. The implication is that *Brahman* is not within time, rather time is within *Brahman*. A second implication is that *Brahman* is the first cause of the universe since, *Brahman* is the only entity that is assumed to be *Ananta,* and therefore, the universe must have come from *Brahman*. Sometimes when we see what is happening on earth, we may feel that it must be the devil who created this universe. Even if it were so, *Brahman* must have created the devil because *Brahman* preceded the devil.

Infinite in objects: This means countably infinite or *Asankhya,* not *Ananta.* Physical objects occupy space and time. If all objects were physical objects, then infinite in terms of space and time would automatically allow for infinite objects. To consider infinite objects as a third *independent* requirement, there must be non-physical objects that do not occupy space and time, such as unmanifest entities. Otherwise, infinite objects will be a corollary of infinite in space and time, not an independent requirement. Infinite in objects means that all objects are in *Brahman* and *Brahman* is in every object.

That *Brahman* is infinite in space, time, and objects means that *Brahman* is here, now, and in every object including you and me. This inner *Brahman* is referred to as *Atman*. Thus, *Atman* is the same as *Brahman*. It means that there is divinity within you and me. We do not have to go looking for *Brahman* outside, instead we must look inward. It means that *Brahman* is not something to be realized after death. *Brahman* could be realized while living.

We shall see later that *Brahman* is homogeneous. Therefore, my *Atman* and your *Atman* are the same. It also means that if I occupy the physical space that you previously did, I now have the *Atman* you previously had. In this sense, our *Atmans* are interchangeable. We are all connected.

Sat

Satyam is the eternal, unchanging reality that underlies all existence. We will abbreviate it as *Sat* and translate it as existence. For something to be *Brahman*, it must also be *Ananta*. What is in every object, that remains unchanged, and is infinite in space and time?

Every object around us is real, every object exists. This laptop exists, this thought exists, this earth exists, and if there is empty space, it also exists. Objects exist, but they change and they are not infinite in space. They were born and will die, so they are not infinite in time. Therefore, objects are not *Brahman*. *This universe is not Brahman.* While objects are not the ultimate reality, they are also not false. A man with a horn or a woman with wings, is false. The objects are neither the ultimate reality nor false, they are temporarily real.

What is common to all objects is the is-ness or existence, which allows the objects to manifest themselves. This laptop is, this thought is, this earth is. This is-ness or existence is the same everywhere. It is homogeneous. Existence has no limit in time, space, and objects.

Existence is unchanged and *Ananta*. Existence is *Sat,* the fundamental essence of all that exists. Existence is *Brahman.*

There are two ways to view the relationship between *Brahman* and objects:

Traditional interpretation: *Brahman* is infinite existence, just as an ocean is a large body of water. Waves arise in the ocean as temporary objects, but waves and the ocean both are water. Similarly, objects temporarily arise in *Brahman* but both are existence. Since *Brahman* is infinite and everywhere, objects cannot arise in *Brahman* as waves do in an ocean – there is air above the ocean for water to take the shape of a wave. This requires a physical transformation, and anything that is infinite and everywhere cannot be physically transformed. Objects arising in *Brahman* are more like an illusion, like mistaking a rope for a snake. *Brahman* is the cause, objects are the effect, but the effect occurs without any change in cause. A more appropriate analogy is that of a stone carver who can see the image he will carve in the stone before any carving takes place. In this interpretation, all objects including physical objects, are the same as *Brahman* but mistaken to be different.

Alternate interpretation: An alternate explanation is as follows. *Brahman* is not existence, but *gives* existence to objects. A close analog is Higgs Boson, also called the God Particle. It is associated with the Higgs field, which permeates all of space. As elementary particles, such as electrons, move through this field, they interact with it. When particles interact with the field, they gain mass. The larger the interaction, the larger the mass. This interaction may be visualized as follows. Think of the Higgs field as a viscous medium that permeates the universe. Particles that interact with the field experience a strong drag from it, appearing heavier. A photon does not interact with this field and is massless. It is not that the Higgs field gives its own mass to particles; it makes mass possible. Similarly, an alternate interpretation is that it is not that *Brahman* gives a portion of Its existence to objects; *Brahman* makes existence possible. In this interpretation, objects are not *Brahman,* but *Brahman* makes their existence possible. We have not pursued this interpretation further in this essay.

Advaita Vedanta (nondual) philosophy accepts the traditional interpretation. That is where the name *Advaita* comes from – that the universe and *Brahman* are not two. The alternate interpretation is akin to the *Sankhya Philosophy* (dual) suggesting that the universe and *Brahman* are two.

Cit

Jnanam is knowledge, made possible by *Cit* – consciousness. We gain knowledge through experiences. Life is a series of experiences. But these experiences are not all-pervasive, eternal, and unlimited. They are not *Ananta* – infinite. Therefore, our experiences cannot be *Brahman*. What is common to all these experiences is consciousness, which allows us to have these personal experiences. I have consciousness, you have consciousness, and so do other living beings. *Advaita Vedanta* says that this consciousness is infinite and same everywhere. It is in living beings and inside a rock, although a rock is unable to demonstrate it, and we are unable to perceive its presence in a rock. Consciousness is *Ananta*. Consciousness is *Brahman*.

Science takes a different point of view. What science calls consciousness is different from what *Advaita Vedanta* calls consciousness. Therefore, we have adopted the nomenclature that when speaking of *Advaita Vedanta,* we use the word consciousness. When speaking of science, we use the word awareness. The scientific argument goes like this: We know that we are aware. We are aware of our surroundings, our thoughts, and feelings, and even aware that we are aware! Awareness varies from person to person and among living beings. It is the current scientific belief that this awareness is an emergent property of the process of biological evolution. First the evolution of living beings and then the evolution of a complex brain driven by the survival arrow of evolution. Science says that matter came first and awareness evolved. Science has yet to understand how this awareness comes about, how we feel first person experiences. It is the hope of science to show how awareness comes as a function of life and complexity.

On the other hand, *Advaita Vedanta* says that consciousness is homogeneous, same everywhere, and is infinite. Consciousness is *Brahman*. Regardless of how objects arise, whether they are an illusion or a result of the process of evolution, it is consciousness that makes awareness possible. Our awareness is conditioned consciousness, namely, consciousness conditioned or limited by our body, mind, intellect, and tendencies. Awareness is caused by the interaction between consciousness and body, mind, intellect, and tendencies. This is like pure water being contaminated by the dirt inside a pot. Consciousness is like pure water. Awareness is like dirty water. Because body, mind, intellect, and tendencies vary from person to person, awareness varies from person to person. Animals have a different degree of awareness than humans because their body, mind, intellect, and tendencies are different from those of humans. Since awareness varies, it is not *Brahman*. Also, we are aware of our awareness. Therefore, our awareness is an object, not a subject. Our awareness is not *Brahman*.

It is possible that how awareness comes about as a function of the brain will be scientifically understood in the future. The current scientific approach appears to be to divide the problem into the hard and the easy problem of awareness. Humans can perform certain functions such as seeing, hearing, speaking, and discriminating. The so called "easy problem" is to explain how these functions are performed, given that we are aware. The so called "hard problem" is to explain why the performance of these functions is accompanied by subjective experience, such as the feeling of pain and pleasure.

However, scientific understanding of the relationship between the brain and awareness will not lead to an understanding of the role of consciousness as defined by *Advaita Vedanta*. It will not resolve the differences between science and *Advaita Vedanta* regarding how awareness comes about. This is because to *experimentally* find the effect of any entity, and to determine the role played by that entity, one must be able to make a change in that entity. For instance, to find the effect of ambient temperature on how a person feels, one needs to change the ambient temperature and observe the change in feelings. Consciousness is homogeneous and unchanged. *Brahman,* as defined by Hinduism, is homogeneous and unchanged. Since it cannot be changed, its effect cannot be experimentally determined. The *Advaita Vedanta Brahman* assumption is not a testable hypothesis that can be experimentally verified.

Corollaries of *Brahman* Assumption

We now show that many words used to describe *Brahman* logically follow from the *Brahman* assumption without having to make additional assumptions.

Ananda

Ananda means bliss – permanent happiness. Note that being happy is the goal of all human beings. It is therefore not surprising that *Brahman* is defined as *SatCitAnanda,* meaning one who makes existence, awareness, and bliss possible. *Brahman* gives existence to all objects, gives awareness to the living, and makes a happy life possible. Previously we have assumed *Brahman* to be *Sat, Cit,* and *Ananta* – infinite. Is *Ananda* (bliss) an additional assumption regarding *Brahman,* or does it follow from the assumption that *Brahman* is *Ananta* – infinite?

Here is one way to practically understand *Ananda.* We all want to be happy. Our experience is that we are happy when our desires are satisfied, and we get the objects we want, be they physical or mental. But our happiness is fleeting. One object may give us happiness for some time, but that feeling disappears, and the desire for another object appears. Permanent happiness lies in satisfying all our desires all the time. But that does not happen, and we do not achieve the state of bliss.

According to *Advaita Vedanta,* we as human beings are not our body, mind, intellect, or tendencies. We are our *Atman,* the same as *Brahman. Brahman* is infinite in objects. Every object is already in *Brahman.* Therefore, the argument goes, if we only realize who we truly are, *Atman-Brahman,* we will realize that we already possess all objects of our desire, and this realization will bring us bliss. *Brahman* is that which makes bliss possible for us.

If *Ananda* is viewed in this manner, it becomes a corollary of *Ananta* because it can be derived from the fact that *Ananta* is infinite in objects. *Sat, Cit,* and *Ananta* are independent of each other, but *Ananda* follows *Ananta* and is therefore not an additional independent assumption. *Brahman* should be called *SatCitAnanta* as the *Taittiriya Upanishad* does, but is often called *SatCitAnanda,* to emphasize that *Brahman* is the source of bliss which we so desire.

Ultimate Reality

Let us now turn to the question of whether *Brahman* is the ultimate reality. *Advaita Vedanta* says so.

What does ultimate reality mean? Reality is something that exists. Objects exist, but they change and are therefore temporarily real. Since we are talking about reality in the context of this universe, we define ultimate reality to be that which is (a) the first cause of the universe, (b) exists eternally, and (c) is unchanged. How many ultimate realities can there be, and what may be their nature? It is possible that there is no ultimate reality – like a continually changing, oscillating universe that has no beginning and end. Otherwise, the number of ultimate realities could be 1, 2, 3, … ∞. Their nature could be matter, consciousness, or something else such as a combination of matter and consciousness.

As one illustration of ultimate reality, suppose it is true that electrons, protons, and neutrons are fundamental particles and are three independent first causes of this universe. All elements that constitute this universe are made by combining different numbers of these particles. Furthermore, let us suppose that these three particles continue to exist eternally, unchanged. If so, then by our definition, they constitute three ultimate realities because they are the first cause of the universe, exist eternally, and are unchanged. Their nature is matter. Note that in this case, ultimate reality (sub-atomic particle) is unchanged but can interact to make different elements.

Science says that this universe came from the Big Bang. If the prevailing conditions at the beginning of the universe no longer exist, then Big Bang may well be the first cause of the universe, but will not be the ultimate reality under our definition if those initial conditions no longer exist.

We now proceed to prove that if *Brahman is the only entity that is Sat, Cit, and Ananta,* then it is the ultimate reality because it meets the three criteria of being the first cause of the universe, existing eternally, and being unchanged.

First cause of the universe: This universe is continually changing, but it does exist. *Brahman* is *Sat,* meaning it can give existence. This universe has entities that are aware. *Brahman* is *Cit,* meaning that it can give awareness. *Sat* and *Cit* mean that *Brahman* can produce a universe that exists and is aware. Furthermore, *Brahman* is *Ananta* – infinite in space, time, and objects. If *Brahman* is the only such entity that is *Ananta,* then infinite in time means that it must be the first cause of this universe because this universe was born a finite time ago. Infinite in time, space, and objects means that *Brahman* is the first cause of everything that there was, is, and will be in this universe. Hence, *Brahman* is the first cause of the universe.

Exists eternally: That *Brahman* exists eternally follows directly from *Brahman* being infinite in time. It was never born and will never die.

Unchanged: Both *Sat* and *Cit,* as defined here, are the same everywhere. They are not to be interpreted as different ingredients of *Brahman* but are different aspects of the homogeneous *Brahman.* Anything that is always homogeneous, infinite, and exists everywhere – inside an atom and in interstellar space – cannot be changed.

Therefore, *Brahman* is the one and only ultimate reality. Its nature is consciousness.

Additional Corollaries

Brahman has been described by different names to communicate the inconceivable. We now show that these descriptions are either inherent or corollaries of the *Brahman* assumption – namely, *Brahman* is *Sat, Cit,* and *Ananta.*

Among the corollaries of this assumption, we have previously shown that since *Brahman* is infinite in space and everywhere, *Brahman* cannot move or has no need to move. We have shown that *Brahman* cannot be cut or bent or changed. We have shown that if *Brahman is Ananta*, then it is also *Ananda* – giver of bliss. We have also shown that *Brahman* is the ultimate reality.

Several stanzas in *Bhagavad Gita* describing *Brahman (Atman)* can be easily understood as either inherent in the definition of *Brahman* or corollaries of the *Brahman* assumption. For example, stanzas 2.20 and 2.23 below.

> *It was never born*
> *Nor does It die*
> *Unborn, eternal, changeless and ancient*
> *It does not die when the body dies.*
>
> *Weapons do not cleave It*
> *Fire does not burn It*
> *Water does not wet It*
> *Nor does the wind dry It.*

Here are some additional descriptions of *Brahman* that also follow from the *Brahman* assumption without having to make additional assumptions.

Ekam – *one and only*. It is our assumption that *Brahman* is the only entity that is infinite and everywhere.

Aja – *unborn*. Our assumption is that *Brahman* is infinite in time. Nothing came before *Brahman*, and it is therefore unborn.

Nirakara – *formless*. An object takes a form when there is something beyond the boundary of the object. *Brahman* is infinite in space. There is nothing beyond it.

Omnipresent – *all pervasive*. It is our assumption that *Brahman* is infinite in space and everywhere.

Nitya *– beyond time.* Our assumption is that *Brahman* is eternal. Time as we know it, began with the beginning of this universe. In this sense, *Brahman* is beyond time.

Nirvikar *– unchanged.* We have shown previously that anything that is infinite, everywhere, and homogeneous will remain unchanged.

Aksharam *– imperishable.* ***Chiranjeevi*** *– permanent.* Both follow from *Brahman* being infinite in time and unchanged.

Nirmala *– stainless.* Because *Brahman* is homogeneous and uncontaminated.

Jnanam *– the knowledge principle.* ***Omniscient*** *– all knowing.* *Brahman* is the knowledge principle by which everything becomes known because *Brahman* is assumed to be *Cit* – consciousness.

Omnipotent *– all powerful.* Because nature came from *Brahman*, nothing can happen in this universe that does not conform to the laws of nature.

Nirguna *– without attributes.* *Sat, Cit* and *Ananda* as well as all the other descriptions of *Brahman* are not properties, qualities, or attributes of *Brahman*. They are the very nature of *Brahman*. *Brahman* is not a mixture of *Sat, Cit* and *Ananda,* rather, the infinite, homogeneous *Brahman* is that which makes existence, awareness, and bliss possible.

Achintya *– inconceivable.* It is difficult to describe existence, consciousness, and infinity.

Atman

The essay "Structure of a Human Being" details the constitution of a human being, including the concept of *Atman*. Here, we proceed as follows.

We begin with the question – Who am I? – making a distinction between who I am (the subject) and what belongs to me (the objects). If I ask, "who are you?" and you respond with your name, profession, or the country of origin, the answers describe your attributes, not who you really are. Similarly, when I say that this laptop is *mine*, I am not identifying myself as the laptop. The laptop is an object; my eyes are the subject. When I say that these are *my* eyes, I am not my eyes. The eyes are objects; my mind is the subject. My thoughts are an object; I am not my thoughts. Anything that I can think about is an object, and not my true self. I am aware of my awareness, so even my feeling of awareness is an object. Who am I, then? I am that which enables this comprehension. It is consciousness which makes these experiences possible. I am that consciousness. This pure consciousness within us is what we call our *Atman*. I am *Atman*.

One cannot comprehend *Atman* because it is the Subject who makes it possible to comprehend objects. A Subject can understand objects, but objects cannot understand the Subject – just as our body cannot understand our mind.

Does the concept of *Atman* follow from the *Brahman* assumption? The answer is yes! *Brahman* is homogeneous existence-consciousness-infinite, which is everywhere including within us. *Atman* is the *Brahman* within us. *Brahman* cannot be divided, but conceptually, *Atman* may be thought of as the portion of *Brahman* within us, finite, uniformly distributed throughout our body, and not just located in our heart or brain, completely connected to the infinite *Brahman*. If *Brahman* were space, *Atman* would be akin to pot-space, the space inside a pot. Space cannot be physically divided, but conceptually, pot-space is a small portion of the total space.

Viewed this way, every human being has an *Atman* that, with our usual ego, we call our own. I think of it as *my Atman*, you think of it as *your Atman*. Sometimes people think that their *Atman* is in their heart or brain. If I move from one room to another, my heart and my brain move with me. But what happens to my *Atman*? Since *Brahman* cannot move – It has no need to move because it already occupies all possible positions – my *Atman* cannot move with me. My *Atman* is not some entity that resides in my heart and moves with it. As we move about, our *Atman* does not move. Our bodies occupy different portions of *Brahman*. Since *Brahman* is homogeneous, our *Atman* remains unchanged. If I occupy the physical space that you previously did, I now have the *Atman* you previously had. This is like saying that if we interchange the location of two pots, each occupies space previously occupied by the other. In this sense, your *Atman* and my *Atman* are interchangeable. We are all connected.

Nondual *Advaita Vedanta,* dualistic *Sankhya Philosophy*, and nondual science interpret *Atman* differently. This is discussed in detail in the essay "The Creation Equation" and summarized below. The concepts of *Maya* and *Prakriti,* referred to below, are also detailed in that essay.

Nondual *Advaita Vedanta* postulates a single ultimate reality, *Brahman,* whose nature is consciousness. *Advaita Vedanta* says that this universe, including humans, is really *Brahman* but is misunderstood to be this material universe due to the projecting and veiling power of *Maya,* the universe-manifesting aspect of *Brahman*. *Brahman* is the cause, this universe is the effect, and *Maya* makes the effect appear without a change in *Brahman*! Therefore, as individual human beings, we are really *Atman* in our entirety, but misunderstood to be this physical body, just as a rope may be mistaken to be a snake. It is not just that *Brahman* is within us, rather, we *are Brahman*. This is true of all objects, physical and mental, animate, and inanimate.

Dualistic *Sankhya Philosophy* classically expresses dualism through two independent ultimate realities: an infinite multiplicity of *Purushas* (individual consciousnesses) and *Prakriti* (primordial matter). For the purposes of this book, and without loss of generality, the infinite *Purushas* are conceptualized collectively as a singular *Brahman*. This allows us to recast the framework into a simplified dualism between *Brahman* and *Prakriti* – each distinct and independently real. The transformations of *Prakriti*, under the presence or backdrop of *Brahman*, are considered to give rise to the manifest universe.

It should be emphasized that the term Dvaita (dualistic) in this context does not correspond to Dvaita Vedanta as established by Madhvacharya, which is rooted in theistic dualism involving a personal God. The discussion of a personal deity is beyond the scope of this work.

Under the dualism of *Brahman* and *Prakriti,* the universe including our physical body, are made of matter and the *Brahman* within our physical body is our *Atman*. Since *Brahman* is homogeneous, all *Atmans* are identical. Our physical body is different from *Brahman* but our *Atman* is the same as *Brahman.*

Science is also *Advaita* (not two), because it postulates one ultimate reality, which is material in nature. Mainstream science says that our awareness is an emergent property of evolutionary biology – a result of the evolution of matter. How this awareness comes about is yet to be understood. Explaining awareness has been divided into two parts: the "easy problem" (how sensory and thought perception occurs, assuming that we are aware) and the "hard problem" (how awareness itself arises, and first-person experiences come about). *Atman,* in the context of science, would be that which makes awareness possible.

Tat Tvam Asi

Tat Tvam Asi means "That Thou Art" or "You Are That." It means that the individual Self *Atman* is the same as the universal reality *Brahman.*

Let us explore how *Tat Tvam Asi* follows from the *Brahman* assumption, in both nondual and dual philosophies.

Nondual Advaita Vedanta

Advaita means not-two, meaning that *Atman* and *Brahman* are not two separate entities. For non-negative numbers, not-two implies any number other than two. Not-two could mean zero, one, or more than two. However, **not-two does not mean many, neither does it mean one, nor does it mean none! The essence of Tat Tvam Asi lies in unraveling this puzzle.**

Advaita Vedanta posits that you are your *Atman*. Since both *Atman* and *Brahman* exist, not-two does not mean none. The *Atman* in all entities is the same because *Atman* is a finite portion of the homogeneous *Brahman* and, as explained below, size does not matter for practical purposes. Therefore, not-two does not mean many or infinite. Are *Atman* and *Brahman* identical? Does not-two mean one?

Conventional interpretation: Are *Atman* and *Brahman* the same? The traditional answer is YES. The philosophy is termed *Advaita* (non-dual) rather than *Ekam* (one) to negate the feeling people have that they are different from *Brahman*.

$$Atman = Brahman \quad \ldots\ldots\ldots\ldots\ldots\ldots\ldots\ldots\ldots\ldots\ldots(2.1)$$

Hence, *Tat Tvam Asi* – That Thou Art! You are your *Atman*, the same as *Brahman*. This is the conventional interpretation of *Tat Tvam Asi*.

Alternate interpretation: Another view is that the equality in Equation (2.1) is not absolute; it is only true for practical purposes because *Brahman* is homogeneous, and the size of *Atman* does not matter for practical living. Hence,

$$Atman = Brahman \text{ (for practical living)} \quad \ldots\ldots\ldots\ldots\ldots(2.2)$$

That size of *Atman* does not matter for practical living can be understood as follows: If we could predict the chemical properties of water by simply knowing the structure of a water molecule, then if a water molecule could talk, it might say to the ocean *Tat Aham Asi* – That I am. This would be true for understanding the chemical properties of water. While many water molecules together make the ocean, a single water molecule cannot produce a Tsunami. An ocean can. A water molecule and an ocean may be the same for certain practical purposes but not in absolute terms.

$$Atman \neq Brahman \text{ (in absolute terms)} \quad \ldots\ldots\ldots\ldots\ldots\ldots(2.3)$$

Such may be the case with *Atman* and *Brahman*. For practical living they are one but not in absolute terms. This could be why the philosophy is called *Advaita* – not-two, rather than *Ekam* – one. Not-two does not mean many or infinite, nor does it mean one in the absolute sense, nor does it mean none!

Dualistic *Sankhya Philosophy*

We are considering *Sankhya Philosophy* with two ultimate realities, *Prakriti*, and *Brahman*. Unlike *Advaita Vedanta*, our physical body is not an illusion here. Our physical body is made of matter, and *Brahman* is consciousness. Matter and consciousness are different from each other and that is why this philosophy is dualistic.

Our *Brahman* assumption is that *Brahman* is homogeneous, and always everywhere. It is uniformly distributed throughout our physical body. If the portion of *Brahman* inside our body is called our *Atman*, then the same arguments apply here as in *Advaita Vedanta*: *Atman* and *Brahman* are the same for practical purposes (*Tat Tvam Asi*) but not in absolute terms.

Mahavakyas

The *Upanishads* contain several *mahavakyas* – great sentences – that explore the concepts of *Brahman* and *Atman*. These include:

1. *PraJnanam Brahman* – "Consciousness is *Brahman*" (*Aitareya Upanishad, Rig Veda*). Consciousness is that which makes all knowledge possible. This is known as a statement of definition.
2. *Tat Tvam Asi* – "That Thou Art" (*Chandogya Upanishad, Sama Veda*). *Tat* meaning That, refers to *Brahman*. *Tvam* meaning Thou, refers to *Atman*. *Asi* meaning art, establishes the oneness of both. The teacher tells the student that you are your *Atman*, and your *Atman* is *Brahman*. This is known as a statement of advice.
3. *Ayam Atma Brahma* – "This *Atman* is *Brahman*" (*Mandukya Upanishad, Atharva Veda*). The student repeats this statement until it becomes the student's realization. This is known as a statement of practice.

4. *Aham Brahmasmi* – "I am *Brahman*" (*Brihadaranyaka Upanishad, Yajur Veda*). The student internalizes what he has been told, practices, and realizes that he is his *Atman,* which is indeed *Brahman*. It is the pronouncement of a Self-realized individual. This is known as a statement of experience.

These four great sentences define *Brahman* as the universal homogeneous consciousness. They tell the student that the consciousness within the student is the same as the universal consciousness. The student repeats the statement and practices until the student realizes its truth.

Practical Significance

The existence of *Brahman* cannot be proved or disproved by conventional scientific methods. *Brahman* is homogeneous, unchanged, infinite, and always present everywhere. Experimental verification of the effect of any entity requires experiments with and without that entity, or experiments with at least some change in that entity. This is not possible because *Brahman* always remains unchanged.

Scientific knowledge about nature, the laws of nature discovered by science, apply to *Maya* or *Prakriti,* not *Brahman*. *Maya* is the veiling and projecting power of *Brahman* that projects this continually changing universe while hiding *Brahman* (see the essay "The Creation Equation" for further details). It is interesting to note that science has made great strides in understanding the universe-generating aspect of *Brahman,* the magic of *Maya*.

Nothing can happen without *Brahman* yet *Brahman* takes no responsibility for anything that happens, good or bad! This can be understood as follows. *Brahman* is like the gas in your car. Gas brings car to life; the car cannot function without it. However, if you drive the car into a ditch, the gas takes no blame for it. Similarly, if you win Daytona 500, the gas takes no credit for it either. Yet these events would not have occurred without gas. Similarly, nothing can happen without *Brahman*, but *Brahman* takes no responsibility for any outcomes.

To win the race, it requires a better car, and a better driver. That is where our focus should be – on improving ourselves and making the world better. That is what is meant by *God helps those who help themselves*.

Swami Vivekananda, a great teacher of Hindu philosophy, elaborates. He provides further guidance on the practical implications of *Tat Tvam Asi* in a beautifully crafted paragraph (paraphrased):

> *"Each soul is <u>potentially</u> divine. The goal is to manifest this divinity within by controlling nature: external and internal. Do this with Karma Yoga, Bhakti Yoga, Raja Yoga, or Jnana Yoga, by one or all – and be free. This is the whole of religion. Doctrines, dogmas, rituals, books, temples, and forms are but secondary details."*

By "soul" is meant *Conditioned Atman* – *Atman* limited by body, mind, intellect, and tendencies. The *Atman* in you is the same as the infinite consciousness *Brahman*. This is a profound confidence booster. However, the *Conditioned Atman* is not fully divine but has the potential to be. The idea that "God is within you" means that we are capable of Godly behavior if we focus on improving our body, mind, intellect, and tendencies. This self-effort will let our divine consciousness exhibit its full potential. Manifesting internal divinity is our personal responsibility.

The same *Atman* is in everyone and everything, whether human or not! We are all potentially divine. The traditional Hindu greeting with folded hands called *Namaste* recognizes this divinity in others. We are all connected by this common divinity. It is as if we are brothers and sisters, connected to the infinite consciousness, each with an umbilical cord called *Atman*. This realization gives us a sense of interconnectedness and unity. It leads to the golden rule – treat others as you would like to be treated – and the idea of *vasudhaiva kutumbakam* – the whole world is my family. It encourages empathy and compassion for others, reminding us that our individual actions affect the collective whole. The ethics of Hinduism are based on this notion of oneness of all.

Our mission is to demonstrate our divinity within by controlling internal and external nature. There are different interpretations of this statement. One interpretation is that control of internal and external nature requires both science and spirituality, as briefly discussed below.

Control of external nature is a matter of science. This is the quest of STEM – science, technology, engineering, and mathematics. Understanding and control of external nature is one aspect of God-like capability. This ability can encourage harmony with nature, promoting ecological consciousness, and responsible stewardship of our planet. It can inspire practices such as sustainable living, conservation, respect for other life forms, and *Ahimsa* – nonviolence.

Equally important is mastering our internal nature – our body, mind, intellect, and tendencies. This is where science and spirituality converge. Control over our inner nature promotes values like – honesty, truthfulness, kindness, integrity, and ethical conduct. This inner transformation empowers us to embody godly values in our everyday lives.

Consequently, we should lead our lives with two primary objectives:

Lead a purposeful life: Strive to make meaningful contributions that leave the world better than it was. This entails setting significant goals and diligently working towards them. It demands courage to persevere through failures, the pursuit of knowledge, and the acquisition of necessary resources. There's also that intangible element – some call it God's grace; others, luck.

Lead a happy Life: Seek personal fulfillment and joy. However, pursuing happiness solely for oneself, without contributing to the well-being of society and environment, is insufficient.

Swami Vivekananda emphasizes that Hinduism offers four paths to achieve these objectives. *Karma Yoga* – The path of selfless action without attachment to outcomes. It is about engaging in deeds that benefit others, cultivating a spirit of service. *Bhakti Yoga* – The path of devotion and unwavering faith in the divine. It involves nurturing a deep, personal relationship with God through love and worship. *Raja Yoga* – The path of meditation and mastery over mind and body. By stilling the restless mind, we perceive the divinity within and attain inner peace. *Jnana Yoga* – The path of knowledge and wisdom. It focuses on intellectual inquiry and understanding the true nature of reality through study and contemplation.

We should focus on the path – or combination of paths – that resonates with our nature. However, these paths are not mutually exclusive; rather, they complement each other, offering a comprehensive approach to spiritual growth.

Furthermore, Swami Vivekananda notes that temples, rituals, books, and related practices, while helpful, are secondary. They are tools that can aid us but should not distract from the primary goal: realizing our inner divinity and living a life aligned with that realization.

Believe in yourself, that is belief in God, because God is within you! Recognizing the divinity within us propels us to move through the world with confidence, making the most of the gifts we have been given. It is essential to develop and implement an action plan aligned with this understanding. Hinduism offers extensive guidance on this journey, encouraging practices like selfless service, and ethical living.

Summary

This essay presents a systematic exploration of Hinduism's (*Advaita Vedanta*) foundational assumption that *Brahman,* defined as existence, consciousness, infinite is the one and only ultimate reality.

Brahman is infinite – in space, time, and objects. This means that *Brahman* is everywhere, eternal, and in all that exists. It means that *Brahman* is here, now, and within you!

Brahman is existence – the is-ness that makes all objects manifest themselves. The essay distinguishes between two interpretations of how objects relate to *Brahman*: the traditional *Advaita* view sees objects as illusory manifestations of *Brahman* Itself, while an alternate interpretation suggests *Brahman* enables existence without Itself being existence.

Brahman is consciousness – that which makes all knowledge possible. Unlike awareness, which science views as an emergent property of biological evolution, *Advaita Vedanta* asserts consciousness as the primary reality from which all experience arises.

From these characteristics, the essay demonstrates that many of the traditional descriptions of *Brahman* logically follow without additional assumptions – including being homogeneous, formless, immovable, omniscient, omnipotent, unborn, the first cause of the universe, and the source of bliss (*Ananda*).

Central to Hindu philosophy is the concept of *Atman* – the individual Self as *Brahman* within each being. Since *Brahman* is homogeneous and omnipresent, all *Atmans* are identical, leading to the profound realization of *Tat Tvam Asi* (That Thou Art) – the individual Self and universal reality are fundamentally the same. This recognition forms the basis for understanding that each person is *potentially divine*, possessing the same infinite consciousness as the ultimate reality.

The practical implications of this philosophy center on manifesting one's inherent divinity through mastering both internal and external nature. External mastery involves scientific understanding and technological capability, while internal mastery requires developing character, tendencies, and behavior. This translates to two primary life objectives: leading a purposeful life that contributes meaningfully to the world, and leading a happy life through personal fulfillment. Hinduism offers four complementary paths to achieve these goals: *Karma Yoga* (selfless action), *Bhakti Yoga* (devotion), *Raja Yoga* (meditation and self-discipline), and *Jnana Yoga* (knowledge and wisdom).

The philosophy's ethical foundation rests on recognizing universal interconnectedness – since we are truly our *Atman,* and the same *Atman* exists in all beings, harming others harms oneself, while serving others serves the divine within. This understanding naturally leads to values like non-violence (*Ahimsa*), compassion, honesty, and the principle of *Vasudhaiva Kutumbakam* (the world is one family). Unlike religions that locate divinity externally, Hinduism's distinctive message is that God is everywhere including within everyone, accessible through spiritual practice and ethical living in this very life. Faith is to believe in yourself, because to believe in yourself is to believe in God. The essay concludes that this represents a radical reimagining of human potential – viewing ourselves not as isolated individuals struggling for survival, but as expressions of infinite consciousness temporarily manifesting to awaken to our true nature, and serve the evolution of all life.

CHAPTER 3
CAUSE AND EFFECT

Are the foundations of Hinduism just figments of poetic imagination – or are they well-grounded in our practical experience?

This essay challenges the common view that Hindu philosophy is mystical speculation. Instead, it reveals a systematic attempt to answer life's deep questions based on observed patterns of causality. This principle of cause and effect that underpins both spiritual thought and scientific inquiry is the second important assumption of Hindu philosophy.

The essay logically classifies our real-world observations of causation and hypothesizes how concepts like karma, reincarnation, Karma Yoga, Brahman, creation, human structure, varna, and Sthitaprajna may have emerged from these experiences.

If you have ever wondered how ancient thought could arise from ordinary experience – or how metaphysics might mirror your own logic – this is the exploration you have been waiting for.

The Causality Assumption

Brahman is the first key assumption of Hinduism. The second key assumption of Hinduism is the causality assumption, the universal law of cause-and-effect, which states that every effect has a cause(s), and every cause produces an effect(s).

Events may appear miraculous, but true miracles – outcomes without causes – do not occur. While we may or may not be able to discover them, there is a deeply held belief that causes do exist, and things do not happen purely by chance or without reason.

Consider the example of John and Peter, who go to fight a war. John returns, but Peter does not. Even though we know that the chance of dying in a war is extremely high, and both could have died, we still seek to understand why Peter did not return while John did. Statistical explanations feel incomplete. Even with a fair coin that produces heads and tails equally over many tosses, we seek deeper causes for each individual result. Our belief is that if we knew the relevant physics and all the necessary information – the weight distribution of the coin, the point of force application, the amount of force, the environmental conditions, the height from which the coin was tossed, and where it was caught – we could predict the result exactly.

Causality is also the bedrock of science. Science holds that everything in the universe, whether physical or psychological, inanimate or animate, will eventually be understood and expressed as mathematical equations representing cause-and-effect relationships in everyday life where uncertainty principle does not disrupt normal experience.

While Hindu philosophy and modern science both use causality to answer fundamental questions, there are major differences. One difference is in the range of questions being addressed. Science addresses both small and large questions, seeking to explain why roses are red and how the universe came about. Hindu philosophy focuses on the large questions. Another difference is that scientific answers are often quantitative, while philosophical answers are usually qualitative. A third difference is that the proof of a scientific theory lies in its predictions being experimentally proven to be correct with 100% confidence. Even a single counterexample can disprove a theory. This means that a scientific theory, while extremely well supported, cannot be proven to be 100% correct. It can only be said that the theory has not been proven wrong. Hindu metaphysics, on the other hand, is usually assumed to be 100% correct, even in the absence of empirical evidence.

Hindu metaphysics presents logical propositions based on cause-and-effect relationships. Modern scientific theories sometimes align with these ancient propositions, and sometimes not.

Practical Causality

Causality operates across all dimensions of human experience. It can be categorized into four main areas: physical, mental, social, and spiritual.

Physical causality involves direct material transformations – exercise strengthens muscles, heat melts ice. These effects are typically measurable and follow discoverable patterns.

Mental causality encompasses how thoughts, emotions, and cognitive processes influence each other – studying improves understanding, stress affects decision-making. These effects are internal and observable through behavior and self-reflection.

Social causality describes how human interactions create ripple effects through communities – economic policies affect employment, cultural movements change social norms. These effects often involve complex feedback loops and emergent properties.

Spiritual causality, as understood in Hindu philosophy, involves practices that influence personal growth and enlightenment – meditation influences awareness, devotional practices change attitudes, ethical behavior influences inner peace. These effects are subjective but form the foundation of spiritual development systems.

These categories of causality reveal consistent patterns in how causality operates in human experience. The law of cause-and-effect, as applied to these categories, can be elaborated as follows:

1. Every effect has at least two courses.
2. Every cause produces effects.
3. Effects follow causes.
4. A cause is inherent in the effect.
5. Some causes are in our control, others are not.
6. The chain of causality is infinite.

7. An effect can appear to occur without a change in the cause.
8. The effect of a cause depends on other causes.
9. The amount of cause may or may not be important.
10. Very few causes produce most of the effect.

1. Every Effect has at least two causes

Can an effect be due to a single cause? Two causes? What is the minimum number of causes necessary to produce an effect in our day-to-day experience?

A common example in Hindu philosophical literature is that of a potter making clay pots out of clay. To make a clay pot, the potter needs:

1. Material Cause: This includes the clay, water, and other ingredients necessary to make the pot.
2. Process Cause: This includes the equipment, such as the spinning wheel and potter's hands, and the necessary energy.
3. Intelligent Cause: This is the knowledge of the type of pot to be made and the process of making it, which comes from the mind and intellect of the potter.

The effect, namely the clay pot, is a modification of the material cause (clay) made using the intelligent and process causes. Such is always the case – that the effect is a modification of the material cause. The material cause may not always be an external physical cause. For mental effects, brain itself is a material cause.

In this example, three causes – material, process, and intelligent causes – are necessary to make a clay pot. Such is the case with all products made by human beings.

Which of the three causes are always necessary to produce an effect?

Material cause: The answer is yes. The alternative is to suggest that something can be created from nothing, which is not our usual experience. This is also the case for natural disasters such as earthquakes—earth is the material cause, and the changes caused by the earthquake are the effects, modifications of the material cause.

Process cause: The answer is again yes. The process includes the energy necessary to effect the change. The process cause is necessary for human-made products and natural phenomena like earthquakes, where there is a process by which they occur, whether we fully understand it or not. Similarly, there is a process by which our thoughts come about and are expressed as writing.

Intelligent cause: The intelligent cause is not always necessary. For example, walking on a sandy beach results in footprints in the sand. Here, the sand is the material cause, and the application of force produced by our weight and the pattern of our walk constitute the process cause. However, there is no intelligent cause unless we deliberately create a specific pattern.

Radioactive materials that emit radiation, such as alpha particles, do so spontaneously. In this case, radioactive material is the material cause, the alpha particles are the effect, and the process of emitting the alpha particles is the process cause. The intelligent cause, as we usually define it, does not seem to exist in this scenario, as well as in the case of creating products from clay without using any intelligence, or in earthquakes.

In usual human experience, particularly with physical causality, an effect comes from the actual transformation of the material cause. The minimum necessary number of causes is two – material cause, acted upon by a process cause, resulting in a transformation of the material cause. The intelligent cause is not always necessary. In human experience, an effect does not come about due to a single cause. It requires at least two causes: a material cause and a process cause (action).

2. Every Cause Produces Effects

A cause can produce multiple effects. For instance, if I push a chair, it moves and scratches the floor. Similarly, if a breadwinner loses his job, the entire family suffers. One person's actions can affect many people, both positively and negatively, producing societal effects. This is particularly true for our interactions with family, friends, and coworkers, but the effects are not limited to those we know. For example, the discovery of the polio vaccine helped millions of people, most of whom were unknown to its discoverer, Jonas Salk.

Our actions produce not only external physical effects but also mental effects, leaving subtle impressions in our minds. If hard work results in success, it may reinforce the belief that hard work is a way to succeed. It is not just our actions but also our thoughts that leave impressions in our mind. Actions taken by other people, including their unspoken thoughts if we can infer them, also leave impressions in our mind. Other actions, including those taken by the rest of the universe, produce impressions in our mind. Examples abound. Praise or insult by another individual, a car accident, an earthquake, a tornado – all produce certain impressions in our mind, whether shallow or deep. Repeated mental impressions result in the formation or modification of certain tendencies, which become second nature. This shows that every cause can produce one or more physical and mental effects.

3. Effects Follow Causes

A cause comes first, and the effect follows. Effects do not have to occur immediately. For example, there can be a considerable time gap between sowing and reaping. An apple seed produces an apple tree, but only after a certain time. Similarly, effects do not have to occur near the cause. If a missile is launched and lands thousands of miles away, the effect is separated from the cause by a large distance. A cellphone invented in the United States provides benefits felt throughout the world. Again, this effect is separated from the cause by space. If a supernova explodes, its effect may not be felt on Earth for millions or even billions of years, separating the effect from the cause not only in space but also in time.

Whether the effect is good or bad may also be understood only later. For instance, artificial sweeteners are a way to avoid the calories from sugar, but if prolonged use results in unanticipated negative effects, they may be discovered much later. An action may produce good effects in one country and bad effects in another, separated by space. Several economic and military actions fall into this category.

4. Cause is Inherent in the Effect

An effect is material cause transformed. The material cause (clay) is present in the effect (clay pot). Oranges come from oranges. Here the orange seed is the material cause that transforms to a tree and eventually produces oranges.

It is often true that good deeds produce good effects, while bad actions produce bad effects. In this case the good deed itself is not getting transformed into good effect in the sense of a physical transformation. The good deed produces an effect on the thought pattern of the recipient, and that in turn leads to external good effects. The brain of the recipient is the material cause, and the effect is a change in brain activity. The effect is still a change in the material cause, except that the material cause that changes is the brain, not the good deed.

While it is true that an effect is a transformation of the material cause, not all transformations are as easy to discern as the clay pot and the clay. For example, when hydrogen and oxygen combine, the result is water. A combustible gas – hydrogen – and a gas necessary for combustion – oxygen – combine to produce water, which can quench fire. By simply looking at liquid water, it is not possible to infer that the material causes are these two gases. It requires chemical analysis to reveal this. This example also shows that the cause being inherent in the effect does not necessarily mean that the properties of the cause and the effect are the same or even similar.

The intelligent cause, which is the intelligence of the potter, is also inherent in the effect but cannot be fully inferred by examining a single pot. This is because the total intelligence of the potter is not used to make any one pot. To infer the intelligent cause, one would have to examine all the creations of the potter, and even then, it may not be possible to fully comprehend the total intelligence of the potter.

The process cause is also inherent in the effect. Can it be inferred by examining the effect? It may be possible to reverse-engineer the process, but not perfectly or always.

That a cause is inherent in the effect may be interpreted to mean that if the cause is removed from the effect, the effect ceases to exist. If the clay is removed from the clay pot, the clay pot ceases to exist. If hydrogen is removed from water, water ceases to exist.

5. Some Causes are in our Control, others are not

Causes may be classified in different ways. Previously we have classified causes as material cause, process cause, and intelligent cause. Here we classify causes into those that are in our control (called control factors) and those that are not in our control (called noise factors). Some control factors can modify the effect of noise factors. In such cases, there is said to be an interaction between the control factor and the noise factor.

For example, suppose we want to administer a therapeutic drug at a constant rate through the skin. There are many advantages to this method of drug delivery, including the need for a lower dosage. Drug transfer is influenced by many causes, some of which are in our control and some are not. Causes such as the amount of drug, surface area of the patch, and thickness of the patch are under the control of the product designer and are called control factors. Human skin also influences the rate of drug transfer. However, skin varies from person to person and from location to location, and it is not within the product designer's control. Skin is a noise factor. In designing such products, if we want to reduce the effect of noise factors on drug delivery, it becomes necessary to find control factors that interact with noise factors – that can reduce the effects of noise factors.

Causes that are not in our control, namely noise factors, come in multiple varieties. Some we cannot control, such as human skin. Other noise factors can be controlled, but doing so is exceedingly difficult or expensive. Examples include ambient temperature and humidity, which in theory could be controlled but would require costly measures like air conditioning.

6. Infinite Chain of Causality

The chain of cause and effect can be infinite. An effect is produced by a cause, which itself is the effect of another cause, and so on. For example, a chair may be made of steel. The steel comes from iron, which comes from taconite pellets, which in turn come from iron ore. The iron in the iron ore was manufactured in the death throes of the first generation of stars, which came from the Big Bang. And where did the Big Bang come from? At the level of detailed processing steps, this is a never-ending chain of cause and effect. Most cause-and-effect chains trace back to an initial cause.

Defining an initial cause can appear as an arbitrary truncation of the causality chain, but that is not necessarily so. This can be understood as follows: imagine a two-dimensional creature crawling on a three-dimensional sphere. The creature is going north, a small step at a time, and after an almost infinite number of steps, it reaches the North Pole. There is no further north to go, and the infinite chain comes to an end at the North Pole. The termination of the cause-and-effect chains at the initial cause could be something similar.

Another approach today is to recognize that modern science suggests that the concepts of space, time, and causation began with the beginning of this universe. The beginning of the universe then becomes the first cause in the infinite chain of causation in this material universe.

If we were to ask, "What existed before the universe began?" there are two possibilities. One is to claim that the question does not make sense because causation began with the universe. The other is to continue the infinite chain until we arrive at a cause that did not need a prior cause to create it.

We have seen before that a cause is inherent in the effect. The implication is that the initial cause is inherent in the universe as we know it.

7. Effect Without Change in Cause

So far, we have observed that in our usual experience of physical causality, an effect is a real transformation of the material cause. An effect occurs due to a change in the material cause. However, is this always the case? Can an effect occur without changing the material cause?

A typical example in Hindu literature is that of a rope and a snake. In twilight, a rope is mistakenly perceived as a snake. Here, the rope is the cause, and the snake is the effect. If the rope is considered the only *material* cause, then remarkably, this effect occurs without any real change in the material cause! Such scenarios often arise from mental misunderstandings or ignorance of the situation. This interpretation will require us to create exceptions to our causality assumption.

Another way to view this is as neurological processes in the brain that create the perception of the snake. The brain is the material cause. The "effect" is not the snake itself, but changes in the brain that create a mental state of perceiving a snake. Thus, the effect is still a transformation of the material cause. It only appears as if an effect has been produced without a change in the cause.

8. Effect Depends on Other Causes

A cause may not always produce the expected effect. For instance, if I push a chair, it may not move if someone else is holding it down. In other words, the effect of a cause or action depends on another cause or other prevailing conditions. Two or more causes may produce an unexpected joint effect. For example, if two forces are applied to a chair, the chair moves not in the direction of any one of the applied forces, but in a different direction determined by the law of the parallelogram of forces.

The following thought experiment involves three medicines – A, B, and C considered as three causes. These medicines are the same but made by different manufacturers. Each medicine is either given to the patient in the recommended dosage or is not given to the patient.

Administering Medicine A to the patient can have different effects as follows:

- Giving only Medicine A improves the patient's condition from bad to good.
- Giving Medicine A when the patient has already taken Medicine B or C makes the patient's condition go from bad to worse due to the side effects of a double dose.
- Giving Medicine A when the patient has already taken both Medicines B and C kills the patient due to the triple dose.

The specific effect depends on whether other medicines were previously given to the patient or not; namely, the effect depends on other conditions or causes. If these other conditions are not known, then the effect can be unpredictable. The same action of administering medicine A could improve one patient and kill another. The effect of a cause depends on other causes.

9. Amount of Cause may be Unimportant

The relationship between cause magnitude and effect magnitude varies. In some physical systems the relationship is proportional – doubling the force doubles acceleration. Sometimes, small causes can produce large effects (a spark igniting a fire). Sometimes large causes produce small effects, such as trying to move an immovable wall. Often, the amount of cause is important. Sometimes it produces unexpected effects. A glass of wine a day may be beneficial, but a bottle of wine a day may not be.

There are situations where the amount of cause is unimportant. While the amount of gas in your car is crucial to determine the distance driven before the tank becomes empty, it does not affect the miles per gallon or the efficiency of the car.

Consider the example of mixing lime juice, sugar, and water to make lemonade. The taste of lemonade depends on the proportion of the three ingredients, not their absolute amounts. The amount of lemonade is important to quench thirst, but not to assess its taste. This shows that in some cases, what matters is the proportion, not the amount.

10. Few Causes Produce most of the Effect

An effect may be due to many different causes, but not all causes are equally important in producing the effect. For example, if I push a chair hard, it moves. However, the force applied to the chair is not the only cause that controls its movement. Other causes include the friction between the chair and the surface it is on, the direction in which the force is applied, environmental conditions, and whether someone is sitting on the chair. The number of such causes is extremely large, even including factors like the phase of the moon, which may have some effect on gravitational force.

Among this multitude of causes, only a few largely control the movement of the chair. While hundreds of factors may cause an effect, only a few causes dominate. This empirical fact is part of the law of cause and effect, known as the effect sparsity principle or the Pareto principle (the 20:80 rule) – 20% of causes produce 80% of the effect.

Consider another example: baking a frozen pizza. Bake temperature and bake time are far more important to the quality of the baked pizza than whether the pizza is on the center rack, whether the fan is on or off, or whether a cookie sheet is used. Similarly, in our lives, what happens to a child is largely controlled by parents, even though it does take a village to raise a child.

Implications toward Hindu Philosophy

Patterns of causality in human experience, discussed above, seem to align with Hindu philosophical thought. Though historical records are insufficient to confirm direct influence, Hindu philosophy appears to be a rational response to universal practical experiences of cause and effect. Exploring these parallels can enrich our understanding of Hindu ideas.

***Karma philosophy*:** Rooted in causality, *karma* theory holds that current circumstances arise from past actions – taken in this life or prior lives – and present actions shape our future. This parallels our everyday experience with cause and effect. Sometimes outcomes defy logic – good actions yield suffering, and bad actions produce good results. This may be because effects depend upon multiple causes and not all causes that produce the effect are necessarily known. Ancient thinkers also proposed that some of these unknown causes may be the result of actions taken in previous lives. This resulted in the theory of reincarnation, suggesting effects may manifest across lifetimes. While unverifiable, this view reinforces causality, and personal accountability. Alternatives – chance or divine predestination – were less favored due to ancient thinker's emphasis on responsibility.

Systemic karma and interdependence: Destiny is defined as what we encounter in life. Practical experience shows that life events are due to multiple causes, some in our control and many beyond our control. Our destiny is shaped considerably by others' actions, whether in this life or their past lives. Innovations like automobiles, global events like wars or economic crises, and even natural disasters shape outcomes for people unconnected to their origin. These systemic influences support the Hindu idea of interconnectedness – *vasudhaiva kutumbakam* (this universe is one family) – and underpin the broader theory of *systemic karma*.

Reincarnation: The cycle of birth and rebirth, driven by residual *karmic* effects, allows cause and effect to stretch across lifetimes. Every cause produces an effect and if the effects are not experienced in this life, there must be next life. Effects may be delayed and unfold far from where causes originated, temporally, or geographically. As in life, not all causes produce immediate or predictable results, reinforcing the logic of reincarnation as a philosophical response to observed uncertainty.

Karma Yoga: Recognizing the practical limits of control over outcomes – results are unpredictable because they depend upon many factors, some in our control and others not – *Karma Yoga* emphasizes action without attachment to results. By focusing on duty over outcome, this path offers a practical and psychological method for sustaining effort despite uncertainty and failure.

Brahman: It is our practical experience that cause and effect can be an unending process. To terminate the infinite process and propose a first cause, *Advaita Vedanta* posits that causality applies only to material phenomena. Consciousness – or *Brahman* – is considered the uncaused cause, and exists beyond material causality. This allows the chain of cause and effect to be terminated in a transcendental source.

Creation: *Advaita Vedanta* views the universe as emanating from *Brahman*, like an effect produced without a change in cause, as in mistaking a rope for a snake in dim light. In contrast, *Sankhya Philosophy* holds that the universe arises from *Prakriti* through material transformations. These reflect our varied everyday experiences of how causality unfolds – mental projection versus material transformation.

Human structure: Hindu philosophy proposes that body, mind, intellect, tendencies, and *Atman* form the human being. These entities are connected by causality. Tendencies lead to thoughts, thoughts filtered by intellect lead to desires, desires lead to actions executed by the body, thoughts and actions influence tendencies, and the cycle continues, explaining how our mental and physical composition evolves through cause and effect.

Chatur Varna: The original *varna* system of Hinduism classified people into four categories based on the idea that an individual's tendencies and actions make them most suited for certain roles and responsibilities in society. Here, tendencies are the cause, and classification, or *varna*, is the effect.

Sthitaprajna: Our peace of mind is disturbed by noise factors – uncontrollable influences like accidents or public opinion. It is our practical experience that factors that are in our control called control factors – mindset, habits, responses – can buffer or eliminate the impact of noise factors. The *Sthitaprajna* exemplifies such control, remaining steady amidst external turbulence.

This analysis of the foundations of Hinduism and the principles of causality suggests that Hindu philosophy is not a collection of random beliefs, but a systematic attempt to address the deepest questions that arise from our practical engagement with life.

Practical Applications

This essay offers several practical applications rooted in understanding causality principles. Here are just a few.

- Our practical experience is that effects are often unpredictable due to unknown causes, and conditions beyond our control. This suggests focusing on performing righteous actions without attachment to specific outcomes as a sustainable way to maintain motivation and ethical behavior even when results disappoint or surprise us.
- We know that factors outside our control can cause emotional disturbances. Our practical experience is that there are factors in our control that can reduce or eliminate the effects of factors that are not in our control. By identifying such factors, we can redesign ourselves to reduce the impact of uncontrollable circumstances on our peace of mind.
- Understanding the interconnected nature of causality – that our destinies are shaped by countless actions from others across time and space – can foster both humility and compassion. This "universe is one family" perspective encourages us to consider the

broader systemic effects of our actions rather than viewing ourselves as isolated actors.
- The Pareto principle (few causes produce most of the effects) suggests focusing energy on identifying and addressing the most influential factors in any situation rather than getting overwhelmed by countless minor causes. This has direct applications in problem-solving, decision-making, and resource allocation.
- Finally, the recognition that effects depend on prevailing conditions encourages more nuanced thinking about why actions succeed or fail, moving beyond simple cause-and-effect assumptions to consider context, timing, and interactions between multiple factors. This can improve both personal decision-making and our understanding of complex situations in relationships, work, and society.

Summary

This exploration reveals a profound truth: the sophisticated philosophical framework of Hinduism may have emerged from humanity's most fundamental observations about causal relationships that make the world work.

The ten principles of practical causality, described in this essay, are not abstract concepts but lived experiences that every human encounters daily. When we trace the logical connections between these universal experiences and Hindu philosophical concepts, a compelling pattern emerges. The unpredictability of outcomes despite our best efforts naturally leads to *Karma Yoga's* emphasis on focusing on righteous action rather than results. The observation that effects of a cause must be experienced and can be widely separated from causes in time and space provides a foundation for reincarnation. The need to find an initial cause in an infinite chain of cause and effect points toward the concept of *Brahman* as the uncaused cause.

Most significantly, this analysis demonstrates that Hindu philosophy is not a collection of mystical beliefs, but a systematic attempt to address life's deepest questions. Whether dealing with personal responsibility, human interconnectedness, or maintaining equanimity amid uncontrollable circumstances, these concepts offer practical wisdom grounded in universal human experience.

The causality assumption thus serves as a bridge connecting ancient wisdom with contemporary relevance. In our modern world, where we still grapple with unpredictable outcomes, complex interdependencies, and the search for meaning amid apparent chaos, these causality-based insights remain remarkably applicable. They suggest that the enduring relevance of these ancient teachings lies not in their historical origins, but in their deep resonance with the universal patterns of human experience – patterns that continue to shape our lives today.

CHAPTER 4

THE CREATION EQUATION

Can this universe be created without depleting its source? This essay presents a simple mathematical equation for the creation of the universe, based on the premise that Brahman is the first cause of the universe and remains unchanged forever. This premise is a corollary of the Brahman assumption. The three solutions to this creation equation reflect the widely different perspectives of Advaita Vedanta, Sankhya Philosophy, and Science regarding the nature and origin of the universe.

In doing so, Sankhya Philosophy introduces the third key assumption of Hinduism: the Prakriti assumption, stating that Prakriti (unmanifest primordial matter) is an independent ultimate reality composed of three gunas (qualities), and this universe is a continuing transformation of Prakriti.

From time immemorial, there has been an insatiable human need to understand how this universe came about. A famous stanza in *Rig Veda* – *Nasadiya Sukta* (Hymn of creation) – composed millennia ago, says:

Who really knows? Who in this universe may declare it? Whence was this creation, who knows whence it arose?

He from whom this creation arose, He may uphold it, or He may not. He who supervises this creation, He knows, or He knows not.

The stanza conveys the belief that creation must have a creator – much like a watch implies a watch maker. It exhibits a sense of wonderment as to who the creator might be, and uncertainty that one may ever be able to find out.

Throughout history, people have pondered questions such as:

1. What is this universe?
2. Who created it?
3. How was it created?
4. When was it created?
5. What will happen to it?
6. Why was it created?

Despite millennia of inquiry, many of these questions remain shrouded in mystery. Science and Hindu philosophy offer different answers to these profound questions.

The non-dual *Advaita Vedanta* says that the universe came from the one and only ultimate reality, *Brahman*. According to this view, the universe is really *Brahman*, though we mistakenly perceive it as a multiplicity of objects.

The dualistic *Sankhya Philosophy* says that there are two ultimate realities, *Prakriti* (primordial matter), and *Brahman* (replaces infinite *Purushas* for this discussion) and that the material universe is a genuine and ongoing transformation of *Prakriti*, facilitated by the interaction between *Prakriti* and *Brahman*.

Science, meanwhile, has yet to settle on a definitive theory of creation. One hypothesis suggests that the universe emerged from "nothing," (quantum vacuum) and consists of continually transforming matter.

These are three among many distinct perspectives on the nature and origin of the universe. We selected these three because their perspectives are so different from each other. Is it possible to reconcile these divergent views? Do they have a common origin? Perhaps these are merely different interpretations of the same underlying reality. Could there be a mathematical equation whose solutions align with the hypotheses proposed by *Advaita Vedanta, Sankhya Philosophy,* and science?

The Creation Equation

We begin by looking at the subject from the *Advaita Vedanta* point of view. The essay "*Brahman, Atman* and *Tat Tvam Asi*" discussed the first major assumption of Hinduism, namely that *Brahman* is *Satyam, Jnanam, and Anantam. Satyam* means existence, *Jnanam* means consciousness, and *Anantam* means infinite in time, space, and objects. Infinite in time means that *Brahman* is the one and only ultimate reality and therefore, must be the first cause of the universe. That essay further concludes that *Brahman* is homogeneous, and because *Brahman* is infinite, everywhere, and homogeneous, it must always remain unchanged. It follows that:

- *Brahman is the one and only ultimate reality.*
- *Brahman is the first cause of the universe.*
- *Brahman remains unchanged (as opposed to unchangeable).*

These are not new assumptions but are corollaries of the *Brahman* assumption.

Today we see a universe consisting of innumerable objects made of matter. The essay "Cause and Effect" discussed the second key assumption of Hinduism: the causality assumption. It was shown that at least two causes are necessary to produce an effect – a material cause and a process cause. It is obvious that this universe must have come from *Brahman* because there was nothing else from which to create it. Even "nothing" did not exist. Hence, the homogeneous *Brahman* is the "material" cause of the universe. We will see later what the "process" cause was. Since *Brahman* remains unchanged, the "material" used during creation was not depleted, unlike the typical human experience where the initial material used in creation diminishes. For example, the more pots the pot maker makes, the less clay remains.

That *Brahman* remains unchanged during the process of creation may be mathematically stated as follows. Let,

x = Initial amount of *Brahman*
y = Amount of *Brahman* used up during the process of creation
(x – y) = Amount of *Brahman* left after the creation of this universe

Since *Brahman* was not depleted in the process of creation, the amount of *Brahman* left after creation must be equal to the initial amount of *Brahman*. Therefore, we have:

The Creation Equation: (x – y) = x ………………… (4.1)

This is the creation equation for our universe. As this is a single equation with two unknowns, x and y, multiple solutions exist, which we hope will correspond to the creation hypotheses of *Advaita Vedanta, Sankhya Philosophy*, and science. To accommodate scientific hypotheses that do not include *Brahman* as the first cause, we generalize the meanings of x and y as follows:

x = Initial amount of material available to create the universe
y = Amount of material used to create the universe

Equation 4.1 continues to apply. There are three important solutions of Equation 4.1.

1. The first solution is (x > 0, y = 0). Since y is the amount of material used during creation, y = 0 means that none of the initial material was used in the process of creation.

2. The second solution is (x = ∞, y > 0). This means that if any portion of initial material was used during creation (y > 0), then the initial material will have to be infinite (x = ∞) for it not to be depleted. This is because infinity remains unchanged if we subtract any finite quantity from it. Also, ∞ - ∞ is undefined but one answer is ∞ - ∞ = ∞.
3. The third solution is (x = 0, y = 0). This means that the initial material available to create this universe was zero (x = 0). Then the material used in the process of creation must also be zero (y = 0). This means that the creation of the universe occurred out of nothing.

These three solutions of the creation equation may be interpreted to respectively correspond to the creation hypotheses of *Advaita Vedanta*, *Sankhya Philosophy*, and science.

From the perspective of *Advaita Vedanta*, x represents *Brahman*, and y is the portion of *Brahman* used to create this universe. The solution (x > 0, y = 0) implies that none of *Brahman* was used during creation, and creation occurred without a real transformation of *Brahman*. This suggests that the effect (this universe) was created without a change in the cause (*Brahman*). If *Brahman* was never transformed, then the universe must be *Brahman*. The multiplicity of physical and mental objects we perceive must merely be a projection or a misunderstanding, like mistaking a rope for a snake in twilight. Hence, *Brahman* and the universe are not two separate entities – this is the essence of *Advaita*, meaning "not two."

Sankhya Philosophy posits two coexisting ultimate realities: *Prakriti* and *Brahman*. (*For our purposes here, maintaining dualism, we have collapsed the innumerable Purushas – consciousnesses of Sankhya Philosophy into a single Brahman*). *Prakriti,* primordial matter, is the material cause of the universe. The ever-changing universe is a continuous transformation of *Prakriti* made possible by the interaction with *Brahman*. There are two scenarios. (a) *Prakriti* gives rise to an oscillating universe where the amount of *Prakriti* used in each cycle is zero.

This corresponds to (x > 0, y = 0) solution. (b) *Prakriti* leads to an expanding universe or an infinity of parallel universes. In this case *Prakriti* will have to be infinite and solution (x = ∞, y > 0) will apply. In either case, according to *Sankhya Philosophy* this universe, being a transformation of *Prakriti*, is composed of matter, while *Brahman* is consciousness. Therefore, *Brahman* and the universe are two distinct entities – hence *Sankhya* is a dualistic philosophy.

From a scientific perspective, one hypothesis proposes that the universe emerged from "nothing" (quantum vacuum), evolving from a zero-energy state. If so, the initial energy (x) available to create the universe must be zero, and the energy used (y) must also be zero. The solution (x = 0, y = 0) applies. The scientific hypothesis further posits that the initial zero energy divided into large amounts of equal positive and negative energies. The positive energy transformed into primordial matter, and the negative energy transformed into gravity leading to our universe.

The three solutions of the creation equation suggest the possibility of harmonizing the three different creation hypotheses of *Advaita Vedanta*, *Sankhya Philosophy,* and science. Let us now delve deeper into these three hypotheses.

Creation Hypothesis of *Advaita Vedanta*

Advaita meaning "not two," suggests that *Brahman* and this universe are not two separate entities.

If *Brahman* is the first cause of the universe and remains unchanged forever, how can the universe be created out of *Brahman* and still leave *Brahman* unchanged? This question is central to the creation hypothesis of *Advaita Vedanta*. The answer that *Advaita Vedanta* gives is that there never was a creation in terms of a real transformation (permanent change) of *Brahman* to matter.

However, the objects in the universe exist. They are born and they die indicating a temporary existence. *Brahman,* on the other hand, exists permanently. *Brahman* was never born and will never die. How is this temporary multiplicity of material reality related to the permanent, unchanging reality?

If *Brahman* is unchanged and this universe came from *Brahman*, then this universe can only be an appearance, a projection, or a misunderstanding – the creation of an effect without a change in cause. A famous analogy in Hindu literature is that of a rope and a snake. In twilight, a rope is mistakenly seen as a snake. The rope is real, and the snake is an appearance. But for the duration of this misunderstanding, the snake is real to us. We are afraid of it. When the truth is realized, the misunderstanding disappears, the reality of the rope is understood, and prevails.

In its appearance as a snake, the rope is not depleted and remains unchanged. Similarly, because there is no actual transformation of *Brahman* into matter in creating the appearance of this universe, y (amount of *Brahman* used in the process of creation) is zero. Thus, the first solution of the creation equation **(x > 0, y = 0)** applies here.

As an aside, it should be noted that Advaita Vedanta says that everything is Brahman. The differences that we perceive, the innumerable objects we see, are due to our ignorance or misunderstanding of the underlying reality of Brahman. Consequently, it is easy to grasp that, potentially, we all are one. The ethics of Advaita Vedanta are derived from this idea of oneness of all. The values we strive for, such as truth, ahimsa (non-violence), and self-control follow from the golden rule – treat others as you would like to be treated, because we are all one.

It should also be noted that the Advaita Vedanta creation hypothesis, by itself, does not require Brahman to be infinite. Just as a finite rope can appear as a finite snake, Brahman could be finite and appear as a finite universe. That Brahman is infinite ($x = \infty$) is part of the Brahman assumption, and is necessary for many reasons, but not essential for the creation hypothesis of Advaita Vedanta.

Maya

The essay "Cause and Effect" showed that at least two causes are necessary for creation – the material cause, and the process cause. *Brahman* is the material cause of this universe, much as the rope is the material cause of the snake. What is the process-cause of the universe? How does *Brahman* project an appearance that we call this universe?

How does our mind project an appearance and create a snake out of a rope, and in doing so, hide the rope? How does the mind create a dream universe? The dreamer experiences happiness and sorrow, meets people, travels widely – all in his dream. For the dreamer, all these experiences are real while they last. In the dream state, our waking state is completely hidden. The mind projects a dream while hiding the reality of the waking state because it is the inherent nature of the mind to do so. Similarly, *Maya,* meaning "that which is not" is the inexplicable power of projection, inherent in *Brahman,* which projects the universe and simultaneously conceals *Brahman.* The objects of the universe are not "created" in the ultimate sense but are experienced due to the mind's superposition. This universe is a result of the interaction between the *Maya-power* of *Brahman* and the mind (also a projection of *Maya*). If the mind were to disappear, *Maya* will not be able to project the universe.

Another way to explain the concept of *Maya* is to consider the question: How many ultimate realities can there be and what is their nature? The number of ultimate realities could range from 0, 1, 2, 3, up to infinity, and their nature could be matter, consciousness, or a combination thereof. If this material universe has existed forever, whether in an oscillating form or as it now appears, then there is no ultimate reality responsible for its creation. The universe always was. In this case, the number of ultimate realities is zero.

Sankhya philosophy posits a material, infinite, ultimate reality called *Prakriti* (see later) and innumerable similar, eternal, independent, and conscious realities called *Purushas*. Here the number of ultimate realities is infinity. Over time, the innumerable similar, conscious *Purushas* were found to be unnecessary and were replaced by a single infinite consciousness called *Brahman*. This resulted in the dualism of *Brahman* and *Prakriti* as the two ultimate realities. *Advaita Vedanta* questioned the need for two infinities, one material and the other spiritual. *Prakriti* was replaced by *Maya,* the universe-manifesting power of *Brahman.* Dualism was replaced by a non-dual philosophy with a single ultimate reality called *Brahman,* endowed with the veiling and projecting power of *Maya.*

Brahman is the material cause of the universe. *Maya* is the process cause that, along with the mind, projects this changing material universe without a change in *Brahman.* This is a situation where an effect occurs without a change in the cause. While the universe as we experience it is not the ultimate reality, the ultimate truth; it is also not false. Something that does not exist at all, such as a unicorn or a ghost, is false. The universe is said to be relatively real, temporarily real, or *mithya* – meaning neither the ultimate reality nor false.

The concept of *mithya* may be further explained as follows. *Advaita Vedanta* recognizes three levels of reality, the lowest level being the dream state (*Pratibhasika* – illusory or dreamlike reality), the middle level is the waking state (*Vyavaharika* – empirical or transactional reality), and highest level is *Brahman* (*Paramarthika* – ultimate reality). The level of reality has to do with dependence. *Brahman* does not depend upon the waking state, while the waking state, being a misunderstanding of *Brahman,* does depend upon *Brahman.* Therefore, it is said that the waking state, namely the universe as we know it, is less real than *Brahman.* The universe is *mithya.*

The above arguments have been beautifully captured in the oft quoted saying:

Brahman satyam, jagat mithya

from *Vivekachudamani,* written by Adi Shankara, the most famous exponent of *Advaita Vedanta* philosophy, who lived in the eighth century, some 1300 years ago. The saying means that *Brahman* is the only truth, and the universe is neither the ultimate reality nor false.

Questions and Answers

Using the philosophy of *Advaita Vedanta,* which posits that there never was a real transformation of *Brahman* to matter, we now address the questions posed at the beginning of this essay:

1. What is this universe? – This universe is the homogeneous ultimate reality, *Brahman*. However, due to our ignorance, we perceive *Brahman* as the innumerable, temporary, and changing physical and mental objects that constitute the universe. Every object is truly *Brahman,* and *Brahman* makes it possible to perceive objects. The following stanza summarizes the *Advaita Vedanta* perspective:

   I am the Subject, not an object,
   Same everywhere.
   And the accomplished know,
   That an object, is also the Subject.

2. Who created it? – It was created by *Maya,* the veiling and universe-manifesting power of *Brahman.*
3. How was it created? – *Brahman* is the material cause of the universe, while *Maya* is the process cause. *Maya* veils *Brahman,* and manifests the universe from *Brahman* without changing *Brahman.* Thus, creation occurs by producing an effect without changing the cause.
4. When was it created? – It is a beginningless universe, like *Brahman* and *Maya.*
5. What will happen to it? – It is an endless universe, like *Brahman* and *Maya.* But, for this ever-changing universe to appear as it does, there needs to be an observer capable of misunderstanding it, and thereby creating it in his mind. If such an observer does not exist, the universe would exist in its true, undistorted state, beyond names and forms.

6. Why was it created? – This question arises due to mind's ignorance. *Brahman* does not act with motive. The universe is not due to a deliberate act but simply the result of *Maya*. However, the universe serves a purpose for us. It is a mistaken appearance, and our purpose is to see beyond this projection, realize the ultimate truth *Brahman,* and achieve liberation.

Creation Hypothesis of *Sankhya Philosophy*

Advaita Vedanta says that *Brahman* and this universe are not two distinct entities – that this universe is truly *Brahman,* but we misunderstand it to be the innumerable material and mental objects. As previously discussed, *Advaita Vedanta* assumes that:

- *Brahman* is the one and only ultimate reality.
- *Brahman* is the first cause of this universe.
- *Brahman* is unchanged forever.

Sankhya Philosophy differs and says that there are infinite ultimate realities – *Prakriti* (primordial matter) and infinite *Purushas* (identical consciousnesses). In our formulation of *Sankhya Philosophy*, we collapse the infinite *Purushas* into a single *Brahman*. Whether we assume infinite identical *Purushas* or a single infinite *Brahman,* it is *Prakriti* and its dynamic interplay of the three *gunas* (see later) that give rise to all forms and experiences. Consciousness – whether plural or singular – does not shape creation. The diversity of the universe emerges from *Prakriti,* not from differences in consciousness.

Under our formulation *Sankhya Philosophy* is a dualistic philosophy and assumes that:
- There are two independent, coexisting, ultimate realities – *Brahman* and *Prakriti.*
- Together, they are two independent causes of this universe. *Brahman* is consciousness, and *Prakriti* is primordial matter. In

contact with *Brahman, Prakriti* continuously transforms to this changing universe.
- *Brahman* remains unchanged forever.

Two important theories of change in *Vedanta* thought are *parinama* and *vivarta*. In the case of *parinama*, the material cause undergoes a real change to produce an effect. The material cause does not remain exactly as it was prior to producing the effect. For example, hydrogen and oxygen combine to produce water. Hydrogen and oxygen are the two causes, and water is the effect. The causes undergo a chemical change at the molecular level to produce water, no longer remaining as they were. Another example is making pots of clay – the lump of clay undergoes a mechanical change in reshaping it as a pot, no longer remaining as a lump of clay. These are examples of *Parinama*.

In the case of *vivarta*, the cause appears to give rise to an effect without a change in the cause. Such is the case when a rope is mistaken for a snake. The rope is the cause, but in producing the effect of a snake, there is absolutely no change in the rope. *Maya* in *Advaita Vedanta* corresponds to *vivarta*, while transformations of *Prakriti* in *Sankhya Philosophy* correspond to *Parinama*.

Although the effect of both *Maya* and *Prakriti* is the same – namely this universe – there is a difference between them. *Maya* is the process cause that produces a mistaken comprehension of *Brahman*, which we call this universe. *Prakriti* is the material cause that continually transforms to produce this universe. Both lead to this material universe. From a scientific viewpoint, understanding the laws governing this universe is to understand the effects of *Maya* and *Prakriti*, not *Brahman*.

Prakriti

Prakriti is the third important assumption of Hinduism, the first two being the *Brahman* and causality assumptions. It is an important assumption of *Sankhya* philosophy. What is *Prakriti*, and where did it come from? We explore two possibilities below – one conventional, and the other a potential alternative.

Conventional interpretation of *Prakriti*: Conventionally, *Prakriti* is taken to be a second infinite, independent, ultimate reality that coexists with *Brahman*. *Prakriti* is an unconscious, subtlest, primordial state of matter. It is nonhomogeneous and composed of three *gunas* (qualities): *sattva, rajas,* and *tamas*. Just as different combinations of the three primary colors result in an infinity of hues, different combinations of the three *gunas* lead to all objects in this universe. Transformations of *Prakriti* bring about insentient objects. Contact of appropriate material objects, called causal and subtle bodies, with *Brahman* is required to bring about conscious, living beings.

Classical texts like the *Bhagavad Gita* 14.10 describe the *gunas* as dynamic and mutually influential, competing for dominance. They do not assign numerical values to *gunas*. To visualize and mathematically model the interplay of the three *gunas,* in this book, we have treated the three *gunas* to add to 100% as a useful teaching and psychological modeling tool (Chapters 7 and 8). It has been implicitly or explicitly used in a similar manner by other contemporary thinkers (e. g. Swami Chinmayananda). Such use is not necessarily scriptural.

There is some similarity between creation from *Prakriti* and scientific thinking. Science asserts that all elements in the periodic table are produced by combining three different sub-atomic particles – protons, electrons, and neutrons in different numbers. For example, hydrogen has one proton, one electron, and no neutrons. Helium has two protons, two electrons and two neutrons. And so on. All objects are made of different combinations of these elements in the periodic table. Creation from *Prakriti* also depends on three *gunas,* but it is the *percentage* of the three *gunas* that matters, not their absolute amounts. Different percentages of the three *gunas* give rise to different objects.

Alternate speculative interpretation of *Prakriti*: The conventional interpretation of *Prakriti* assumes it to be a second independent ultimate reality, thus violating the assumption that *Brahman* is the one and only ultimate reality. We now propose an alternate speculative interpretation that allows *Sankhya Philosophy to* be based on the same *Brahman* assumption as *Advaita Vedanta*. *This interpretation assumes* *that Prakriti is not an independent ultimate reality but is created from Brahman by a continuous or periodic actual transformation.*

Perhaps this interpretation was not conventionally proposed because it appears to involve a change in *Brahman,* while our assumption is that *Brahman* remains unchanged. *However, unchanged does not necessarily mean unchangeable.* If *Brahman* is infinite, a transformation of *Brahman* to *Prakriti* can occur without any change in *Brahman* (infinity minus something finite or even something infinite can equal infinity). Under this interpretation, where *Prakriti* is created from *Brahman, Prakriti* is not a second independent ultimate reality. *Brahman* is the only ultimate reality, is the first cause of the universe, and remains unchanged forever. These are the assumptions made by *Advaita Vedanta. This interpretation allows Sankhya Philosophy to be based on the same Brahman assumption as Advaita Vedanta while permitting it to be dualistic.*

This interpretation says that some unknown portion (y) of *Brahman* transforms to *Prakriti* without a change in *Brahman*. This is creation of matter from consciousness. It is an actual transformation of *Brahman,* namely *Parinama* and not *vivarta*. <u>This is parinama without a change in cause</u>*!* *Brahman* is the material cause of this universe. *Brahman* is also the process cause that transforms a portion of *Brahman* while leaving *Brahman* unchanged.

How can such a process be visualized? A mental picture of such a process may be like that of Hilbert's paradox of the infinite hotel. David Hilbert was one of the most influential mathematicians of the last century. The process of creation may be as follows: *Brahman* is homogeneous, infinite, and everywhere. Imagine it divided into cubes (not actually divided, just in imagination). Let us say that cube one is transformed into *Prakriti*. Simultaneously, cube two is moved to cube one, cube three is moved to cube two, and so on. Even with cube one transformed, *Brahman* continues to be infinite and everywhere. <u>*This is a change in Brahman that leaves Brahman unchanged*</u>*!* A similar argument applies to transforming a finite number of cubes. What if an infinity of cubes were transformed? The infinite even numbered cubes (2,4,6....) could be transformed, and the infinite odd numbered cubes could be simultaneously moved appropriately to leave *Brahman* unchanged. In all these instances, *Brahman* is transformed to matter but remains unchanged and cannot be said to be broken up into parts. *Brahman* is changeable yet unchanged!

This interpretation suggests that *Brahman* is both the material cause and the process cause of this universe. What about the intelligent cause? It is not possible to infer whether *Brahman* is also the intelligent cause, even based on a complete understanding of this universe – just as whether the potter used intelligence to create the pots cannot be inferred by looking at the pots.

Note that this *Sankhya Philosophy* interpretation of the creation hypothesis requires *Brahman* to be infinite. Such was not the case for the *Advaita Vedanta* interpretation. Also, unlike the case of *Advaita Vedanta,* this *Sankhya Philosophy* interpretation requires the additional assumption that *Brahman* can be spontaneously transformed into *Prakriti*. *Brahman* is consciousness. The properties of consciousness are likely to be different from the properties of matter. It may well be that consciousness is self-actuating and can simultaneously be the material cause and the process cause of the universe.

Creation under *Sankhya Philosophy*: In terms of the creation of the universe, there are two scenarios.

- One possibility is that *Prakriti* gives rise to an oscillating universe. The universe expands, then the rate of expansion slows, eventually becomes negative, the universe starts contracting and goes back to its original state. Over the course of one cycle, the amount of *Prakriti* used is zero. This corresponds to **($x > 0$, $y = 0$)** solution, where y is the amount of *Prakriti* used per cycle.
- The second possibility is that *Prakriti* leads to an expanding universe or an infinity of parallel universes. In this case *Prakriti* will have to be infinite and solution **($x = \infty$, $y > 0$)** will apply. It allows *Prakriti* to produce an infinity of parallel universes.

Questions and Answers

We now answer the questions posed at the beginning of this essay under the conventional dualistic interpretation that *Prakriti* is primordial matter, and is a second independent reality.

1. What is this universe? – This universe is made of matter and is a real and continuing transformation of *Prakriti*. This changing material universe and unchanging *Brahman* coexist as separate entities.
2. Who created it? – *Brahman* and *Prakriti* cooperate to produce this universe.
3. How was it created? – *Prakriti* is the material cause, and *Brahman* is the process cause of the universe. In a manner of speaking, the insentient *Prakriti* can move but is blind, while *Brahman* can see but cannot move. Through cooperation, *Brahman* guides *Prakriti,* and *Prakriti* transforms to produce this universe.
4. When was it created? – It is a universe without a beginning – like *Prakriti* and *Brahman*.
5. What will happen to it? – It is a never-ending universe, like *Prakriti* and *Brahman*. The transformations of *Prakriti* will continue whether there are observers to observe them or not.
6. Why was it created? – The universe was not created as a deliberate act but arises naturally from the interaction between *Brahman* and *Prakriti*. The interplay of consciousness and matter serves as a means for self-realization.

Creation Hypothesis of Science

Note that in the following discussion, there are intriguing conceptual similarities between certain physics principles and these philosophical traditions, but we are not suggesting that ancient metaphysical concepts anticipated modern scientific discoveries in literal ways.

One scientific school of thought suggests that this universe originated from "nothing" and involves a continuous transformation of matter. It thus includes features of both the non-dual *Advaita,* and the dualistic *Sankhya* philosophies, namely, y = 0 (*Advaita*) coupled with real transformations of matter (*Sankhya*).

Einstein's general theory of relativity predicted an expanding universe. This prediction was experimentally confirmed by the observation that galaxies are rapidly moving away from each other, like spots on an expanding balloon. An expanding universe means that it must have started from a point source at some time in the past. Modern estimates suggest that the universe began approximately 13.8 billion years ago, in what is called the Big Bang. Despite the name, the beginning of the universe was not an explosion of matter but rather an expansion of space.

Where did this universe come from? The universe consists of a large amount of matter, such as our solar system and galaxies. It may also have dark matter and dark energy that science still does not know much about. All of this adds up to a large amount of positive energy. However, the universe also has gravity. This gravity counts as negative energy. One current estimate is that that the sum of the positive energy of matter, and the negative energy of gravity is zero. This means that the total energy of the universe is zero. In other words, it took no energy to create the universe, making it the ultimate free lunch. The universe came from "nothing."

This corresponds to the (x = 0, y = 0) solution of the creation equation 4.1, meaning that there was zero total energy to begin with (x = 0) and zero energy was needed to create this universe (y = 0). *Advaita Vedanta* says that this universe is the result of *Maya* – no transformation of *Brahman* occurred. Hence, for both science and *Advaita Vedanta*, y = 0.

It is often said that "something" cannot come from "nothing." If "nothing" is really "something," then clearly "something" can come out of "nothing." Also, if the cause of "nothing" is "something," (being a part of a system) then "something" can come from "nothing." For example, suppose my bank account has zero balance. It has "nothing" in it. The bank account had one thousand dollars in it at one time, but a friend needed money, and I gave him a 1000-dollar loan. At that point, the account had "nothing" in it. When he returns the loan there will be one thousand dollars or "something" in the account. In this way, "something" can come out of "nothing" when the original "nothing" came from "something." The implication is that if this universe came from "nothing" then that is possible if (a) "nothing" has "something" in it or (b) if "nothing" came from "something." In either case, it means that there must have been "something" before the big bang.

Quantum field theory tells us that even a so-called empty vacuum is teaming with matter and anti-matter particles in equal proportion. This is the precursor to the Big Bang. These particles are constantly being created, only to disappear shortly afterwards. This creation occurs out of an empty vacuum - out of "nothing." This may mean that empty vacuum has zero energy, but the empty vacuum is not "nothing," it is "something." Our universe originates when, from emptiness, a mass of potential particles pop into existence. The zero energy gets divided into a large amount of positive energy and an equal amount of negative energy. The positive energy transforms to matter, and the negative energy becomes gravitational energy. The sum of the two energies continues to be zero. This feature of creation of the universe from unmanifest energy is like the transformation of *Prakriti* in the case of *Sankhya Philosophy* from an unmanifest state to a manifest state. *Thus, the scientific hypothesis of creation incorporates elements from both Advaita Vedanta and Sankhya Philosophy.*

It is the quest of science to discover the laws of physics to explain how the universe initially came about, transformed itself into what we see today, and what will happen to it in the future. This quest includes explaining not only the multitude of transformations of matter but also of the creation of living, conscious beings including humans capable of asking and answering existential questions.

What was there before the Big Bang? Space, time, and causality originated with the Big Bang. Did the laws of physics also originate with the Big Bang? Perhaps they did, for this universe – but did they always exist in a more general form? If they always existed, those laws of physics are the equivalent of the universe generating aspect of *Brahman* from a scientific point of view.

The conventional scientific explanation assumes that the Big Bang resulted in primordial matter, which, through the process of evolution, has produced this material universe. This leaves open the question of explaining how consciousness came about. The current scientific position is that consciousness is an emergent property of evolutionary biology.

Alternate interpretation
An alternate scientific explanation is that consciousness is not *Brahman* as assumed by *Advaita Vedanta,* and not an emergent property of biological evolution as assumed by science. Rather, consciousness is an inherent property of primordial matter that resulted from Big Bang. One possibility is that elementary particles, such as atoms, may have properties of both matter and consciousness. This will make explaining consciousness easier.

For illustrative purposes, such particles may be pictorially represented by the symbol that appears on the front cover of this book – OM inside a circle with a dot. OM represents consciousness and circle with a dot, resembling an orbiting electron symbolizes matter. This reflects the view that consciousness is an intrinsic and fundamental aspect of all matter, as also suggested by Panpsychism.

Many other interpretations of this symbol are described in the front of this book.

Questions and Answers
We now answer the questions posed at the beginning of this essay based on the standard scientific interpretation.
1. What is this universe? – This universe consists of innumerable objects that exist temporarily in time and space. Each object is a continuing transformation of matter, though not the same transformation. What we perceive as the universe is not the ultimate reality. A table that appears smooth from a distance reveals

scratches upon closer inspection, appears different under a microscope, and is composed of atoms. Atoms are made of electrons, protons, and neutrons, which in turn are made of quarks, and so on. Perception and reality differ; that is *Maya* or the effect of *Maya* or at least the magic of *Maya*. It is interesting that *Maya* should be understood as laws of physics. Modern science suggests that it is the conscious observer who brings objects into existence, while *Advaita Vedanta* says that the conscious observer and the observed are the same.

2. Who created it? – The laws of physics created the universe. This is like the *Advaita Vedanta* statement that *Maya* created the universe.

3. How was it created? – The universe came from "nothing." Note that "nothing" does not mean the literal absence of everything. "Nothing" is teaming with virtual particles that undergo existence and annihilation. Creation from nothing implies ($x = 0$, $y = 0$) or creation from a zero-energy state using zero energy. For both science and *Advaita Vedanta*, $y = 0$; however, for science, it means creation from zero energy, while for *Advaita Vedanta*, it means that the universe is a projection without a change in *Brahman*.

4. When was it created? – It began about 13.8 billion years ago with the Big Bang.

5. What will happen to it? – This question remains unanswered. The universe may continue to expand, or the expansion may slow, leading to a contraction and ending in a big crunch, resulting in an oscillating universe.

6. Why was it created? – One explanation is that it is in the nature of the laws of physics to create universes, just as the laws of evolutionary biology do not necessarily require intent. Universes are being created all the time. Some universes may die early, while others may live long lives. Not all universes will have the right values of universal constants to create entities

capable of asking, in the Nobel prize winning physicist Richard Feynman's words:

I wonder why?
I wonder why?
I wonder,
Why I wonder?

We have come far in answering these questions and yet not far enough. The sense of wonderment expressed in *Nasadiya Sukta* continues.

Practical Applications

The creation equation and its solutions provide a neutral language for comparing creation narratives across traditions. Religious leaders, philosophers, and educators can use the three solutions to facilitate meaningful conversations between seemingly incompatible worldviews. Rather than debate which perspective is "correct," the equation demonstrates how different traditions might be addressing the same fundamental mystery through different conceptual lenses. ***The mystery common to all three perspectives is that the universe was created by a process that did not deplete its source.***

Teachers can use the creation equation to introduce students to the relationship between mathematics and metaphysics, and the evolution of scientific thought. The framework helps students understand how ancient philosophies and modern science often grapple with similar questions using different methodologies, fostering both scientific literacy and cultural understanding.

This unified approach suggests that humanity's diverse attempts to understand existence may be complementary rather than contradictory.

Summary

This essay proposes that a simple creation equation $(x - y) = x$ can reconcile the widely different creation hypotheses, **from Brahman to Big Bang,** proposed by *Advaita Vedanta, Sankhya Philosophy*, and modern science. Here x represents the initial amount of material available to create the universe, and y represents the amount used in creation. The equation's fundamental premise is that the ultimate reality remains unchanged during the process of creation. The equation has three meaningful solutions that correspond to the different philosophical and scientific perspectives.

The first solution, $(x > 0, y = 0)$, aligns with *Advaita Vedanta's* nondual philosophy, where *Brahman* is the sole ultimate reality and the universe is merely an appearance created by *Maya,* the projecting power of *Brahman*. No part of *Brahman* was consumed in creation.

The second solution, $(x = \infty, y > 0)$, corresponds to *Sankhya Philosophy's* dualistic framework, which with our assumption of innumerable *Purushas* being replace by *Brahman,* posits two coexisting ultimate realities: *Brahman* (consciousness) and *Prakriti* (primordial matter). Here, the universe results from actual transformation of infinite *Prakriti* guided by *Brahman's* consciousness, allowing real transformation of matter while ensuring that the infinite source remains undiminished.

The third solution, $(x = 0, y = 0)$, reflects the scientific hypothesis that the universe emerged from quantum vacuum – from "nothing" – through the division of zero total energy into equal amounts of positive energy (matter) and negative energy (gravity). This parallels both *Advaita Vedanta* concept of creation without depletion and *Sankhya's* emphasis on material transformation.

The essay demonstrates how these disparate traditions address the same fundamental mystery: **how can creation occur without depleting its source?** While *Maya* produces apparent transformation without real change, and *Prakriti* undergoes actual transformation while maintaining an infinite source, modern science proposes creation from a zero-energy state that requires no net energy expenditure. The essay offers a mathematical framework that harmonizes these different creation hypotheses, providing a neutral language for comparing creation narratives across traditions. The essay acknowledges that the profound sense of wonder about existence – beautifully expressed in the ancient *Rig Vedic* hymn *Nasadiya Sukta* – remains as compelling today as it was millennia ago, reminding us that the ultimate questions about the nature, origin, and purpose of the universe continue to challenge human understanding across cultures and centuries.

CHAPTER 5
STRUCTURE OF A HUMAN BEING

Ever wondered what truly makes you "you"? This essay takes you far beyond the skin and brain – into a hierarchical structure of a human being as envisioned by Hindu philosophers: gross body, subtle body, causal body, and Atman. This framework – the fourth key assumption of Hinduism – is a bold alternative to materialist views, offering a lens through which reincarnation, karma, and liberation become realities. If you have ever felt like there is more to life than molecules and neurons, this journey into the fourfold human structure may forever change how you see yourself.

Human Structure

Hindu philosophers describe a human being as consisting of a gross body, subtle body, causal body, and *Atman*. The gross body is what we normally refer to as our inert physical body and includes our senses, organs of action, and our brain. The subtle body consists of our mind and intellect, functioning as the seat of thoughts and decision-making. The causal body may be thought of as being composed of three *gunas* (qualities) called *sattva* (balance), *rajas* (activity), and *tamas* (inertia) that drive our *vasanas* (tendencies). The causal and subtle bodies are also the repository of the accumulated effects of our past actions. The causal body reflects our deeper nature and predispositions. Finally, *Atman* is the *Brahman* within us.

Figure 5.1 illustrates this structure with *Atman* at the top, followed by the causal body (showing the three gunas denoted by S, R, and T), then the subtle body (mind and intellect), and finally the gross body at the base. The feedback loop on the left shows how experiences modify the causal body composition over time. It is a hierarchical structure as suggested by *Bhagavad Gita* 3.42.

The senses are superior to dull matter,
Mind is higher than the senses.
Intellect is still higher than the mind,
The Atman is even higher than the intellect

Figure 5.1: The Structure of a Human Being

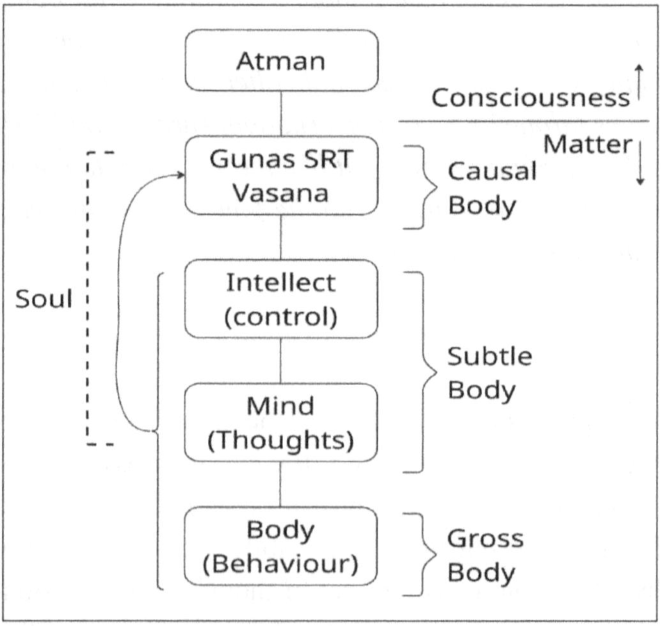

This hierarchical structure of a human being can be justified from multiple perspectives.

Functional perspective: The hierarchy reflects functional control and dependency. Atman enables all experience but remains unaffected. The intellect directs and evaluates but cannot function without consciousness. The mind generates thoughts and emotions but is influenced by *vasanas* and requires intellectual guidance. The gross body executes actions but needs mental direction.

Seer – seen perspective: The hierarchy reflects the *Vedantic* principle of *Drik Drishya Viveka* - the discrimination between seer and seen. The fundamental rule is that the seer and seen are always different. A table is seen by the eyes, the eyes are seen by the mind, the mind is seen by the intellect, the intellect is seen by the witness consciousness (*Atman*). Since the seer is always superior to the seen, and *Atman* is the ultimate seer that can never become an object of perception, it stands at the apex of the hierarchy.

Cause-and-effect perspective: Starting from the top, *Atman* makes all cause and effect possible. Our tendencies (*vasanas*) and our intellect are the causes, and the thoughts in our mind are the effects. Our thoughts, modulated by our intellect, lead to our desires. Our desires drive actions executed by the gross body. This cause-and-effect sequence outlines the hierarchy depicted in Figure 5.1.

Happiness perspective: Senses give us physical pleasure, the most basic form of happiness. Mind provides emotional satisfaction, superior to physical pleasure. Intellect brings intellectual satisfaction, which is higher still. Tendencies satisfy our inherent nature. Being able to identify with the *Atman* gives us freedom from suffering, granting permanent happiness, bliss.

Evolutionary perspective: As expressed by the 13th- century Sufi poet Rumi, the human hierarchy reflects an evolutionary journey from dull matter to life, from life to mind, from mind to intellect, and finally to *Atman* (Nicholson, R. A. (1914). *The mystics of Islam*, Methuen & Co.). Rumi eloquently describes this progression:

I died as a mineral and became a plant.
I died as a plant and became an animal.
I died as an animal and became a man.
I shall die as a man to soar with angels.

As shown in Figure 5.1, apart from the *Atman*, which is the same for all and is non-material, all other entities – gross, subtle, and causal bodies – are made of inert matter. These entities vary for everyone, making each one of the eight billion of us unique. A right contact with *Atman* brings the insentient matter to life, just as the right contact between the gas and the engine brings a car to life.

An analogy: The human structure can be compared to a computer system, where different layers correspond to the gross body, subtle body, causal body, and *Atman*. The gross body is akin to the hardware of a computer – made of physical materials but incapable of functioning independently. Just as a computer cannot operate without software, the gross body remains lifeless without the underlying forces that animate it. The subtle and causal bodies together resemble software and firmware, controlling the functions of the system. The subtle body acts like an operating system, managing sensory experiences, thoughts, and emotions, while the causal body serves as firmware or deep-rooted code, storing *gunas* that shape the individual's tendencies, and karmic patterns that shape destiny. However, none of this works without electricity, which represents *Atman* – the pure consciousness that powers everything. Without *Atman*, the physical, mental, and karmic layers cease to function, just as a computer without power is utterly useless.

This analogy illustrates dependent relationships with the limitation that consciousness (*Atman*) is fundamentally non-material, unlike electricity which is a form of energy.

Gross Body

The gross body is the physical, tangible form that we perceive directly through our senses. It is composed of matter and remains inert without the life force, serving as a mere vehicle for experience. It is an instrument through which perception and action take place. If neglected – whether through misuse, improper nourishment, lack of exercise, or dismissing it as inconsequential – it deteriorates, affecting both its physical condition and functional capabilities.

The gross body functions as an input-output mechanism, facilitating interaction with the external world. The organs of perception – eyes, ears, nose, tongue, and skin – allow the individual to experience colors, sounds, smells, tastes, and touch. Meanwhile, the organs of action – speech, hands, feet, excretion, and reproduction – enable expression, movement, and essential bodily functions, all coordinated through the brain. Impairment in these mechanisms can lead to distorted perception and ineffective action; for example, mishearing someone may result in incorrect responses, just as weakened limbs may hinder mobility.

The gross body is like computer hardware – made of physical materials but incapable of functioning independently.

Subtle Body

The subtle body, or *sukshma sharira*, includes mind, intellect, memory, ego, and vital energies. It is composed of subtle matter. While undetectable by our gross senses, it remains material, and can therefore interact with the physical body. It is responsible for mental activity, and the life force that sustains us. Through this subtle body, we experience thoughts, emotions, and sensations, and make decisions shaping our interactions with the external world.

An integral aspect of the subtle body is its connection to the physical sense organs. It incorporates *subtle senses* that work in tandem with our gross senses – ears, skin, eyes, tongue, and nose. This interaction gives rise to the five types of sensations: sound, touch, sight, taste, and smell. Awareness is a prerequisite for us to perceive and experience these sensations.

The subtle body is active in waking, and dream states. It is also active in transition between lifetimes. The significance of the subtle body becomes particularly evident in the dream state. While the gross body rests, the subtle body remains active, allowing perception, thought, and experience within the dream world. Even though the physical sense organs are inactive, their subtle counterparts continue functioning, creating dream sensations that feel real. This highlights the subtle body's ability to operate independently of the gross body.

The subtle body is also the carrier and storehouse of *prarabdha karma, agami karma* (see Chapter 7) and mental impressions.

The subtle body is like the operating system of a computer - invisible to direct observation but governing all functions, responding to inputs, and maintaining continuity between sessions.

Mind

Our mind is a subtle body – made of subtle matter, but cannot be detected by touch, smell, sight, taste, and hearing. Our five subtle senses give rise to the mind. By virtue of being material, a subtle mind can interact with an appropriate gross body, much like an electromagnetic field, although subtle, interacts with certain physical objects. However, the precise laws governing these interactions remain a subject of discovery.

The mind, enlivened by consciousness and functioning through the brain, is the seat of thoughts, feelings, and desires. The objects of the world that we perceive give rise to a continuous flow of thoughts in our mind. The same object may produce widely different thoughts and desires in different people and in the same person at different times. The translation of perception to thoughts and desires is influenced by our inherent tendencies, likes, and dislikes. If our mind feels happy, we are happy; otherwise not.

Our mind has the weakness that it is never steady. It fluctuates between worrying about the future, and fretting over the past – rarely resting in the present. This causes stress. The present is where the action is and where the focus should be.

To understand the mind in greater depth, we must explore the concept of *antahkarana*, or the "inner instrument" of perception, which encompasses the subtle body. The *antahkarana* operates through four distinct functions:

1. *Mana* (Mind): Collects sense impressions from the five senses forming undecided thoughts.
2. *Buddhi* (Intellect): Serves the evaluation and decision-making function.
3. *Ahankara* (Ego): The "I-thought" giving rise to a sense of I-ness and personal experience
4. *Chitta* (Memory): Acts as a reservoir of memory and impressions.

For our purposes here, we have distinguished between mind and intellect, with mind encompassing everything except the intellect.

Intellect

Intellect, like the mind, is a subtle body but serves a distinct purpose. It is also made of subtle matter, remains imperceptible to human senses, and interacts with the mind. The intellect plays a pivotal role in guiding and regulating our thoughts and desires, acting as the decision-making faculty within us. Unlike the mind which continuously generates thoughts, it is through the intellect that we reason, contemplate, judge, and decide. It also encompasses our free will, empowering us to make deliberate choices.

Our thoughts and desires drive our actions, but it is the intellect that exercises control, preventing us from acting on every fleeting thought that arises in the mind. Acting as a filter, the intellect evaluates the thoughts generated by the mind, discards the irrelevant or unhelpful, directs the mind to produce new thoughts, and chooses among alternatives to form actionable desires. A well-functioning intellect is essential for navigating the world effectively, serving as the internal compass that aligns our actions with wisdom and purpose.

Intellect vs. intelligence: It is important to distinguish between intellect and intelligence. Intelligence refers to the accumulation of information from various sources, such as books, the internet, education, and lived experience. It equips a person with knowledge in specific fields. For instance, technologies like AI (e.g., ChatGPT) possess vast intelligence, often exceeding human knowledge in certain areas. Similarly, institutions like the CIA (Central Intelligence Agency) are repositories of intelligence about global affairs, not repositories of "intellect."

Greater intelligence does not necessarily indicate superior intellect. Intellect involves the ability to apply reasoning and self-regulation to knowledge and thoughts. An individual with vast intelligence may still engage in unwise or unethical behavior if their intellect is weak. For example, a scientist might succumb to sensory pleasures or addictions despite their extensive knowledge.

When the intellect is overpowered by the mind, self-control diminishes, and actions are driven by unchecked impulses rather than reasoned judgment. This distinction between intellect and intelligence is crucial to understand that true success in life is not merely about acquiring knowledge but about applying it wisely, ethically, and with foresight.

Free-will: In making decisions, intellect is often presented with multiple options. For instance, if I wish to travel from Minneapolis to Chicago, four hundred miles away, I could choose to fly, take a bus, drive, cycle, or even walk. The choice depends on what is feasible and appropriate under the given circumstances. This ability to select from various possibilities is what we call *free will*.

However, this raises a fundamental question: Does intellect truly possess free will, or is it merely an illusion? Are the decisions made by the intellect genuinely autonomous or are they predetermined by prior conditions, and physical laws?

A belief in free will provides a sense of control, alleviating feelings of helplessness. This perception is beneficial for mental well-being. It encourages individuals to function as though their choices matter. It fosters personal responsibility, motivating better behavior, and discouraging indulgence in harmful actions.

If free will were an illusion, it would imply that our actions are entirely dictated by physical laws and past conditions. If choices were entirely predetermined, the concepts of personal accountability, praise, blame, or punishment would collapse. The foundations of legal, ethical, and moral systems, as well as metaphysical frameworks like *karma*, reincarnation, and liberation in Hindu philosophy, would lose their validity. Preservation of Hindu metaphysics is not a reason to validate free will, and the question remains: Are there reasons to believe that free will exists?

Can we identify free will through observation? If free will were absent, human choices should be fully predictable under all circumstances, as long as the governing laws and initial conditions are deterministic and known. For instance, I can predict with certainty what a chair will do under specific conditions, which suggests that it lacks free will.

What about a coin toss? While I cannot predict whether it will land heads or tails, this unpredictability does not imply free will. Given perfect knowledge of the coin's weight distribution, the applied force, the environmental factors, and all relevant physics, the outcome could be predicted. What appears as free will is, in such cases, simply complexity coupled with incomplete information about initial conditions.

Humans are exponentially more complex than coin toss and have many more initial conditions. As with the coin, free will may be said to be an illusion, caused by complexity and lack of information if the laws and the initial conditions governing humans are entirely deterministic.

Free will requires unpredictability. Modern science, such as the uncertainty principle, suggests that predictions cannot be made with 100% certainty. Hence, one condition for free will is scientifically met. Even so, it does not mean that the final decision is being made by the intellect. The decision could just be a random outcome. Unpredictability does not prove free will but it does not negate it either. The question then becomes: what distinguishes purposeful choice from mere randomness?

In Hindu philosophy, free will emerges from the intellect's capacity for discrimination. Unlike the mechanical responses of matter or the instinctual reactions of animals, humans possess intellect – the faculty to evaluate, discriminate, and choose between options based on wisdom rather than mere conditioning. While *vasanas* create tendencies, the intellect's discriminative power allows humans to transcend their conditioning. This is not randomness but conscious choice guided by wisdom.

Let us proceed with the assumption that free will exists. Two of the factors that shape human desires and actions are:

1. *Vasanas*: Deep-seated tendencies and inclinations shaped by past experiences, that stem from the *guna* composition of the causal body.
2. Free Will: The conscious exercise of choice by the intellect, which allows individuals to shape their future despite their predispositions.

Causal Body

A causal body is the source of an individual's core tendencies. It consists of the subtlest form of matter in unmanifest (potential) form, not detectable by usual scientific means. Every individual possesses a unique causal body (*karana sharira*), composed of a distinct proportion of the three *gunas* – *sattva* (purity), *rajas* (activity), and *tamas* (inertia). The three *gunas* are like the three primary colors. Just as by mixing the three primary colors in different proportions, we can make an infinity of hues; similarly, by changing the proportion of the three *gunas,* an infinity of human tendencies and personalities come about.

A causal body is also the repository of *sanchit karma*. It is said to be active in deep sleep and in transition between lifetimes.

A casual body is like a deep-rooted code, storing gunas and karma.

Vasana

Vasanas are our innate tendencies or inherent nature emerging from the *guna* composition of the causal body. They govern our thoughts, desires, and actions, often without conscious awareness, and serve as a lens through which we experience the world. These experiences modify *vasanas,* gradually altering our underlying *guna* proportions, as shown by the feedback loop in Figure 5.1.

The manifestation of *vasanas* is dictated by the *guna* composition:

- *Sattvic vasanas* foster harmony, wisdom, and virtuous behavior. They manifest as pleasant actions. A person who selflessly helps others without external motivation may be guided by these elevated tendencies.
- *Rajasic vasanas,* driven by passion and restlessness, often fuel ambition, desire, and impulsivity – leading to both constructive and destructive actions.
- *Tamasic vasanas,* rooted in inertia and ignorance, manifest as indifference, delusion, or destructive tendencies, obstructing clarity and growth and resulting in unpleasant behavior.

While *vasanas* influence our thoughts and behavior, humans possess free will – the capacity to consciously control desires and think and act independently of these tendencies. Free will allows individuals to rise above their *vasanas,* and shape their destinies through deliberate and mindful choices.

Gunas

Sankhya philosophers describe *Prakriti* (primordial matter) as consisting of three *gunas – sattva, rajas,* and *tamas.* These are often understood as inherent qualities or attributes of *Prakriti,* but they can also be seen as its fundamental constituents. They compete for dominance (*Bhagavad Gita* 14.10). As stated earlier, in this book, we are treating this idea of dominance by assuming that the three *gunas* add to 100%. This is a useful teaching and psychological profiling tool. Their percentages constitute our causal body and their interplay governs our *vasanas* (tendencies) and influences our thoughts, desires, and actions.

1. *Sattva* embodies purity, harmony, and balance. It is associated with clarity, knowledge, peace, and lightness. A *sattvic* state

fosters wisdom, calmness, equanimity, and spiritual growth. For example, a person performing an act of kindness selflessly operates in a *sattvic* state. *Sattvic* life is a life of poised, mature, and contemplative achievements.
2. *Rajas* signifies energy, passion, and activity. It drives action, desire, and ambition but often leads to restlessness and overexertion. A *rajasic* person might strive for wealth or fame, sacrificing peace of mind in the process. *Rajasic* life is a life of selfish actions and involvement in the world.
3. *Tamas* represents inertia, ignorance, and resistance to change. It manifests as lethargy, confusion, and stagnation. While *tamas* can facilitate rest and regeneration, excessive *tamas* may result in harmful habits and procrastination. The life of a *tamasic* person is dull and inactive.

It is the proportion (or percentage), not the quantity, of the three *gunas* that matters. They exist in varying proportions within all things, shaping the diversity of phenomena in the universe. Every human being has a causal body with a unique proportion of the three *gunas*. The *gunas* are not static; the proportion (percentage) of each *guna* varies from time to time and topic to topic, so that if an individual is *sattvic*, it is so *on average*. The proportion of *gunas* can be changed through deliberate practice. Meditation and self-discipline, for example, can elevate *sattva*, reducing the influence of *rajas* and *tamas*. These three percentages, which add to 100%, determine our tendencies and have a large influence on our thoughts, desires, and actions.

Life is a learning process. We can change our personality by learning. In responding to our destiny, *vasanas* produce thoughts in our mind. Intellect examines these thoughts and decides using our free-will, turning thoughts into desires. Desires produce action. An examination of the consequences of our thoughts, desires, and actions leads to learning. This learning influences the makeup of the causal body by changing our *vasanas* and the proportion of the three *gunas*. The cycle continues, as shown by the feedback loop in Figure 5.1. When a person dies, he has a different causal body compared to what he was born with.

Atman

At the top of the hierarchy is *Atman,* the eternal Subject, the possessor of all that we refer to as "mine" – body, mind, intellect, tendencies, and even awareness. These are objects of possession, but the *Subject, the one who possesses, is Atman.* This is captured succinctly in the *Vedantic* assertion, "I am, therefore I think," challenging Descartes' proposition, "I think, therefore I am." The act of thinking is not the proof of *Atman*; rather, it is *Atman* that makes thinking possible.

Atman is the *Brahman* within us. It is often described as the unmanifest consciousness, distinct from matter, and the force that enables matter to act. Unlike matter, which requires external energy and various causes – material, process, and intelligent – to produce an effect, consciousness provides its own causes and energy. This interaction between consciousness and matter allows matter to act, making *Atman* the substratum for all existence and awareness. How consciousness interfaces with matter is a deep mystery. It may be that consciousness functions like the Higgs field – making awareness possible when matter has the right configuration.

Nothing can happen without the *Atman;* and yet *Atman* takes no responsibility for whatever happens! This may be understood as follows: *Atman* is like the gas in your car. Nothing can happen without the gas. However, if you drive the car into a ditch, who is at fault? It is certainly not the fault of the gas in your car. Yet this event could not have happened without the gas in your car. If you win the Daytona 500, who gets the credit? Once again, it is not the gas in your car. Yet you could not have won the race without the gas in your car. Nothing can happen without *Atman*, but *Atman* neither takes the blame nor the credit for anything that happens!

Life comes about because of the contact between causal/subtle body and our gross body in the presence of our ever-present *Atman*. This contact makes us aware or conscious. When this contact between causal/subtle body and the gross body, is lost, death occurs.

Alternate interpretation of *Atman*: The accepted Hindu view is that *Brahman* is infinite consciousness and existence. *Atman* is the consciousness within us which makes life and awareness possible.

There is a potentially alternate view that *Brahman* is not necessarily consciousness but something that makes awareness and life possible. This may be understood as follows by analogy with Higgs boson, also called the God Particle.

Higgs boson is associated with the Higgs field which permeates all of space. As elementary particles, such as electrons, move through this field, they interact with the field. When particles interact with the field, the resistance appears as mass. The larger the interaction, the larger the mass. If particles do not interact with the field, they are massless.

This interaction may be visualized as follows. Think of the Higgs field as a viscous medium that permeates the universe. Particles that interact with the field experience a strong drag from the field, appearing heavier. A photon does not interact with this field and is massless. *It is not that the Higgs field gives its own mass to particles; it makes mass possible.* Similarly, it may be that *Atman/Brahman* is not existence and consciousness, but makes existence and awareness possible. This is not a widely accepted interpretation, but provides an intriguing perspective that aligns with modern physics. We have not pursued this interpretation further in this book.

Atman is like electricity without which computer hardware and software do not function, except that Atman is consciousness and electricity is energy.

Soul

Soul, known in Sanskrit as *Jiva*, refers specifically to conditioned *Atman*, while *Atman* refers to unconditioned consciousness. Soul is *Atman* conditioned or limited by the insentient causal and subtle bodies. Put another way, causal and subtle bodies enlivened by *Atman* constitute *Jiva* – the individual soul as shown in Figure 5.1. It is *Atman* that brings life and awareness to these bodies, enabling all experience, perception, and action. Acting as the knower, the doer, and the agent of experience, *Jiva's* existence is veiled by ignorance – an illusion that obscures the understanding that we are, in fact, the unconditioned *Atman*. Liberation lies in discarding this veil and realizing our true essence.

The *Bhagavad Gita*, in Chapter 14, Verse 5, vividly describes this conditioning through the influence of the *gunas* – *sattva, rajas,* and *tamas* – which bind the *Atman* to the body and mind.

> *Sattva, rajas and tamas*
> *Qualities of matter*
> *Bind the immortal Atman*
> *And the body together.*

Swami Vivekananda's assertion that "Each soul is potentially divine" further elaborates on this concept. The soul, while not inherently divine, has the potential to become so under the right circumstances. This journey toward divinity requires recognizing and transcending the limitations imposed by the causal and subtle bodies.

The concept of conditioning may be further understood as follows. Suppose the same gas can be used in the engines of a lawnmower, a car, and an airplane. The performance of the gas is partly determined by the power of the engine. A lawnmower moves slowly, a car drives fast, and a plane can fly. The performance of the gas is limited or conditioned by the specific engine it occupies. *Atman* is similarly conditioned or limited by the causal and subtle bodies it interacts with. *The conditioned Atman is the Jiva* – the individual soul. This perspective illustrates the interplay between *Atman* and the individualized experience of *Jiva*.

Chariot Analogy

The *Katha Upanishad* beautifully illustrates the human structure using a chariot analogy. *Atman* is the passenger, the silent observer who makes the ride possible. Body is the chariot. Intellect serves as the charioteer. Mind acts like the reins. Horses represent senses.

Key insights from the analogy: The *Atman*, while vital for enabling the journey, remains inactive, observing silently. The body, senses, mind, and intellect take center stage in driving the chariot. The intellect as the charioteer, exercises control – determining speed, navigating obstacles, and charting the path forward. The senses, depicted as horses, require discipline; unchecked, they veer off course and the reins, representing the mind, can go haywire. The mind communicates sensory input to the intellect and executes control actions to guide the senses.

Practically, this analogy emphasizes the importance of intellectual discrimination. A wandering mind and uncontrolled senses lead to chaos, but a disciplined mind and senses – like well-trained horses – enable the charioteer to steer the chariot successfully toward the destination.

Extending the analogy to include causal body: The analogy from *Katha Upanishad* gains further depth when we introduce the causal body and *vasanas*. The causal body functions like a businessperson, unseen yet pivotal, negotiates the terms of the journey with the *Atman* and selects the chariot based on his *vasanas*. Through successive rides, the businessperson learns – refining his *vasanas* based on experiences and outcomes until he learns to make each ride perfect.

Each ride symbolizes a lifetime, forming part of the Hindu philosophy of reincarnation. Upon death, the causal body – bearing the three *gunas* (*Sattva, Rajas, Tamas*) – alongside the subtle body transmigrates, taking rebirth. Destiny arises from past *karmas*, while self-effort determines how one responds to these conditions. This dynamic interplay reshapes the causal body and *vasanas* over successive lifetimes.

The goal is transcendence: as the *vasanas* transcend *gunas*, one becomes *jivanmukta* (liberated being).

Human Personality

Figure 5.2, popularized by Swami Chinmayananda, shows how human structure influences human personality.

Figure 5.2: Human Personality

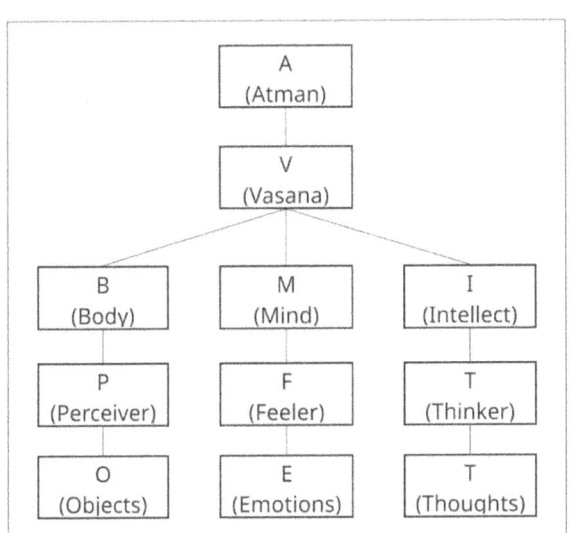

Atman enlivens the body to perceive physical objects of the world using the sense organs. This makes the human being a perceiver. The individual becomes an actor when he acts through the organs of action (not shown in Figure 5.2). Together these roles constitute the physical personality of the human being.

Atman functioning through the mind creates the feeler, who experiences the world of emotions such as love, sorrow, happiness, and anger. This forms our emotional personality.

Atman functioning through the intellect manifests the thinker – the evaluator of thoughts regarding the world. It not only evaluates these thoughts and makes decisions but also contemplates higher truths, forming both our intellectual personality and spiritual personality.

Our perceptions, feelings, and thinking depend upon our *vasanas*, our inherent tendencies. Body, mind, and intellect, governed by *vasanas*, experience the world of objects, emotions, and thoughts, thus forming our composite personality as perceiver, feeler, and thinker. We mistake ourselves as our personality and believe that – I perceive, I feel, I think.

In this process *Atman* becomes conditioned or limited by our body, mind, intellect, and *vasanas* resulting in what is known as conditioned consciousness, or the soul.

Life and death

Life – *Atman*, functioning through the causal/subtle body, enlivens the gross body, making life possible. While *Atman* is omnipresent and always in contact with the causal/subtle body, life emerges only when the causal/subtle body establishes the right connection with the gross body.

To further illustrate: even a rock possesses *Atman* – the *Brahman* within it. However, *Atman* cannot bring a rock to life because a rock lacks the causal/subtle body necessary to establish this connection.

Death – Medical definitions of death have evolved from cessation of breathing and heartbeat to brain death. In the context of *Advaita Vedanta*, death occurs when the connection between the gross body and the causal/subtle body is severed. Even in death, the immovable *Atman* remains within the dead body. It is the causal/subtle body that transmigrates, setting the stage for rebirth in another gross body – a process called reincarnation.

Consider an analogy. Gasoline powers a car by interacting with its engine, which creates the requisite connection between the gas and the car's physical components. Gas cannot animate a rock or even a human because they do not have the proper mechanism for connection. This reinforces the idea that the right interface, like the causal/subtle body, is vital for life.

Scientific View

Current scientific understanding of human structure is different from the above. Science today takes a purely materialistic approach to human structure and one scientific view is that life force and consciousness are emergent properties of evolutionary biology, and will be understood as functions of gross body. There are other hypotheses such as consciousness being an intrinsic property of matter.

Over centuries, medical science has made groundbreaking strides in understanding the physical – or gross – body. Heart transplants can now save lives, genetic engineering addresses hereditary defects, and neuroscience delves into the functioning of the brain.

However, progress remains limited when it comes to consciousness, and what Hinduism refers to as the subtle body, and causal body. The current scientific discourse categorizes the challenges of understanding consciousness into two distinct realms: the easy problem and the hard problem of consciousness. Ironically, these "easy" problems are anything but simple, having confounded researchers for decades.

Easy problem: The easy problem asks: *"Given that we are aware, how is the content of our awareness generated?"* This realm includes questions that, in principle, are addressable via computational or neural mechanisms. It examines:

- Sensory processes: How do we see, hear, smell, taste, and feel?
- Brain functions: How is sensory data processed, how are thoughts formed, and how are decisions made?
- Behavioral consequences: How do thoughts and decisions shape actions?

In Hinduism terms, the easy problem corresponds to understanding the interaction of the causal and subtle bodies with the gross body, for example, understanding the role of subtle senses. Scientifically, it is to understand the role of brain and neurons in generating sensory inputs, thoughts, and decisions. They are called easy problems because they appear to be amenable to be solved by the usual scientific methods.

Hard problem: The hard problem ventures deeper. *"How does subjective experience – awareness – arise?"* It explores why sensory functions are accompanied by personal, qualitative experiences. A self-driving car, for instance, can perform tasks such as seeing, hearing, and decision-making, yet lacks subjective experience. The hard problem seeks to explain how personal experience manifests in humans – a quest analogous to explaining the soul (*Jiva*) in Hindu philosophy.

In Hinduism, the hard problem involves understanding how *Atman* renders the causal and subtle bodies conscious leading to subjective experiences. In both scientific and philosophical contexts, consciousness is a property not typically associated with matter, making it an elusive and profoundly challenging inquiry. Hence the term "hard problem."

The articulation of the distinction between the easy and hard problems of consciousness appears to be recent. Yet Hindu philosophers, millennia ago, conceptualized a similar framework, distinguishing between causal/subtle matter and the soul, enlivened by *Atman*.

Scientific notion of consciousness – let us call it *awareness* – differs fundamentally from Hinduism's concept of *Brahman* (pure consciousness). The effect of *Brahman* being homogeneous, infinite, and everywhere cannot be understood conventionally. Therefore, for Hinduism, the hard problem cannot be solved. Science on the other hand believes that awareness is an emergent property of evolution. Scientific solution of the hard problem involves explaining how awareness evolved.

This may be further understood as follows. Imagine a world permeated by an invisible gas that powers vehicles. In this world, a car with an engine is "alive," while a stone or human is "dead" without an engine. Scientific analysis may conclude that the engine is the source of life, without grasping the invisible gas's role. Similarly, science may identify mechanisms of awareness without perceiving the role of *Atman*, the ultimate enlivening principle, if the Hindu idea of consciousness is correct.

Hinduism postulates a hierarchical framework for human existence: the gross body, subtle body, causal body, and *Atman*. This forms the foundation for metaphysical concepts such as *karma*, reincarnation, and liberation. This hierarchical framework is the fourth key assumption of Hinduism, following the *Brahman*, causality, and *Prakriti* assumptions.

Practical Applications

Understanding the hierarchical structure of human beings offers practical insights for personal development and daily living. By recognizing that our physical health directly impacts our ability to perceive and act in the world, we can prioritize proper nutrition, exercise, and rest as foundational practices that support all higher functions. The distinction between intelligence and intellect becomes particularly valuable in managing stress and making decisions. When we understand that the mind generates continuous thoughts while the intellect serves as the discriminating faculty, we can develop practices like meditation and mindfulness to strengthen our intellectual capacity to evaluate and filter mental activity rather than being overwhelmed by every passing thought or emotion.

The concept of *vasanas* provides a framework for understanding why we have certain behavioral patterns and tendencies while empowering us to consciously work on transforming them. Rather than feeling helpless about ingrained habits, we can recognize these as expressions of our *guna* composition that can be gradually shifted through deliberate practice. Cultivating *sattva* through activities like contemplation, service, and ethical behavior naturally reduces the influence of *rajas* (restlessness) and *tamas* (inertia) in our lives. This understanding also helps us approach personal growth with patience, recognizing that fundamental change occurs at the causal body level, and requires sustained effort over time.

In relationships and communication, this framework encourages us to look beyond surface-level personality differences and recognize that everyone operates from their unique combination of *vasanas* and *guna* proportions. This perspective fosters greater compassion and reduces judgment, as we understand that others' actions stem from their conditioning rather than inherent character flaws.

The chariot analogy particularly emphasizes the importance of developing intellectual discrimination as the charioteer that can guide our senses and mind toward beneficial outcomes. It suggests that strengthening our capacity for wise decision-making is one of the most practical investments we can make in our overall well-being and effectiveness in life.

Most profoundly, understanding *Atman* offers a radical shift in self-identity that can transform how we experience life itself. By recognizing ourselves as the unchanging observer rather than the constantly changing thoughts, emotions, and circumstances, we can reduce identification with temporary states. This perspective suggests that true fulfillment comes not from perfecting our personality or circumstances, but from recognizing our essential nature as pure consciousness. *When we understand that "I am, therefore I think" rather than "I think, therefore I am," we can approach daily challenges from a place of fundamental stability and peace, knowing that our core identity remains untouched by success or failure, pleasure or pain.*

Summary

This essay presents the Hindu philosophical understanding of human structure as a hierarchical system consisting of four components: gross body, subtle body, causal body, and *Atman*. The gross body represents our physical form – the tangible matter we can perceive through our senses, including organs and brain. The subtle body, composed of subtle matter beyond sensory detection, encompasses our subtle senses, mind, and intellect, functioning as the seat of thoughts, emotions, and decision-making. The mind generates continuous thoughts from perceptions, while the intellect serves as the discriminating faculty that evaluates these thoughts and exercises free will in making choices.

The causal body, the subtlest form of matter, consists of three *gunas* (qualities) – *sattva* (balance), *rajas* (activity), and *tamas* (inertia) – that shape our *vasanas* or inherent tendencies. These *vasanas*, accumulated from past experiences, influence our thoughts and behaviors, though humans retain free will to transcend these predispositions. The causal body also stores the effects of past actions (*karma*). At the apex of this hierarchy sits *Atman*, described as pure consciousness or *Brahman* within us – the eternal observer that enables all awareness and experience while remaining unaffected by actions or their consequences.

The essay illustrates this structure through the *Katha Upanishad's* chariot analogy, where *Atman* is the passenger, the body is the chariot, intellect is the charioteer, mind represents the reins, and senses are the horses. Life emerges when the causal and subtle bodies establish proper connection with the gross body. Death occurs when this connection is severed, with the causal and subtle bodies transmigrating to enable reincarnation. The framework contrasts with current scientific materialism, which attempts to explain consciousness as an emergent property through the "easy problem" (how awareness content is generated) and the "hard problem" (how subjective experience arises) classifications. Hindu philosophy maintains that consciousness (*Atman*) is the fundamental reality that makes all experience possible rather than being an emergent property of matter.

Finally, the essay presents practical applications toward daily living, personal development, relationship management, and leading a happy, productive life.

CHAPTER 6

THEORY OF REINCARNATION

What if the story of your life did not begin at conception – or end at death? Hindu philosophy proposes a breathtaking answer: you are a soul on a timeless journey, born repeatedly, shaped not just by genes and upbringing, but by ancient tendencies and actions forged across lifetimes. This essay takes you deep into the theory of reincarnation – a fundamental assumption of Hindu philosophy – comparing it with current scientific views including Darwin's theory of evolution.

Whether you are seeking meaning in adversity, hope beyond mortality, or a richer understanding of human development, this journey invites you to rethink what it means to suffer, grow, evolve – and truly live.

Hinduism does not believe that death signifies the ultimate end of a person. Unlike the notion that "from dust we come and to dust we shall return," Hinduism asserts that individuals are born again and again. Upon death, the soul transmigrates and takes rebirth in a new physical body. This process, known as reincarnation, is the fifth key assumption of Hindu philosophy, following the *Brahman*, causality, *Prakriti*, and human structure assumptions.

This essay presents Hinduism's theory of reincarnation and explains how it relates to previously discussed concepts. It examines how reincarnation theory explains individual differences, suffering, and spiritual development, and compares this explanation with scientific materialism. Their differing explanations of the birth of a baby illuminate deeper philosophical differences about agency and responsibility. The essay compares the mechanisms of spiritual evolution (through *karma* and rebirth) with biological evolution (through natural selection), and shows surprising structural similarity regarding gradual development through challenge-response processes over long time periods.

Life after Death
According to Hindu philosophy, the life we are leading right now is, in fact, life after death. We were born before, died before, and are now born again in our present incarnation. To see life after death, one simply needs to look in the mirror. We see reincarnated beings all around us. Viewed in this manner, reincarnation loses some of its mystique. Hindu philosophers further suggest that human life is the highest form of existence, and if we live it wisely and ethically, we shall return as human beings.

Bhagavad Gita
As described previously, *Bhagavad Gita* is one of the three pillars of Hindu philosophy. It presents Krishna's teachings to the warrior Prince Arjuna on the eve of the Kurukshetra war. In the opening verses (Chapter 1 and the first ten verses of Chapter 2), Arjuna expresses deep sorrow and inner conflict. His hesitation does not stem from a fear of dying or killing – he understands his duty to lead his army and fight in a just war. Rather, his dilemma arises from his attachment to his family, friends, and teachers, who now stand as his opponents. Struggling to reconcile his role as a warrior with the moral weight of battling his own kin, he leans toward the belief that attachment surpasses duty.

Recognizing Arjuna's turmoil, Lord Krishna begins his *discourse from the chariot* by introducing reincarnation. Had Krishna merely functioned as a consultant, he might have quickly explained *Karma Yoga* – the path of selfless action. Instead, he chooses to be a true teacher, delivering a profound and comprehensive exploration of Hindu philosophy over the eighteen chapters of the *Bhagavad Gita*.

Krishna first addresses Arjuna's fundamental misconception – his attachment to the physical body. He redirects Arjuna's focus from the transient nature of the body to the eternal essence of the soul, explaining that the soul never dies but simply transitions from one body to another. Only after laying this foundation does Krishna introduce *Karma Yoga*, teaching action performed without attachment to outcomes. By beginning with reincarnation, Krishna affirms its significance within Hindu philosophy, opening his discourse with these words from Chapter 2, verse 12:

> *There never was a time,*
> *When you and I did not exist*
> *Nor will there be time,*
> *When the living will not be*

And then again, a little later, Chapter 2, verse 22:

> *Just as one replaces,*
> *Used garment with a new one,*
> *So does the Self discard,*
> *Needless body for another*

These well-known verses in the *Bhagavad Gita* suggest reincarnation – a cycle of birth and death that continues eternally. Krishna concludes his advice in the *Bhagavad Gita* by saying, in Chapter 18, verse 63:

> *I have taught you,*
> *The most secret sacred knowledge*
> *Reflect on it fully,*
> *Then do as you will.*

Great teachers will be disappointed if the students followed what they heard merely because the teacher said so. Therefore, Krishna ends *Bhagavad Gita* by telling the common person – *Think and then decide what you choose to believe!*

Let us think about this theory of reincarnation.

Reincarnation

Reincarnation is the fifth fundamental assumption of Hindu philosophy. It asserts that upon death, the soul – consisting of causal and subtle bodies enlivened by *Atman* – transmigrates and takes rebirth in a new physical body. This cycle continues till the soul attains freedom from rebirth.

The theory of re-incarnation is based upon the previous four assumptions of Hindu philosophy, coupled with the reincarnation assumption stated above. The relevant aspects of the four assumptions are stated below.

1. *Brahman,* being infinite and all-pervasive, cannot move – it has no need to move because it already occupies all positions. *Atman,* the presence of *Brahman* within us, also remains motionless.
2. The effects of one's actions must be experienced.
3. The causal body contains unique proportion of the three *gunas* – *sattva, rajas, tamas* – and shapes an individual's tendencies.
4. The soul consists of causal and subtle bodies enlivened by the *Atman*.

Theory

Life is a series of experiences that begin with conception and birth, continue throughout our existence, and end with physical death. The circumstances we encounter, known as destiny, are shaped by our actions in this life and actions in previous lives whose effects we are to experience in this life because they were not experienced in pervious lives, necessitating rebirth. Both favorable and unfavorable circumstances are part of our destiny. Our effort in dealing with our destiny provides us with learning opportunities. This learning process does not end with physical death; rather, the individual is born repeatedly until the individual learns to embody divine qualities, achieving *Moksha* – liberation from the cycle of life and death – or becoming a *Jivanmukta* – free from suffering in this life.

Our next life depends on how we manage our destiny through actions taken in this and previous lives. It is not always easy to understand why specific experiences come our way. This can be compared to the application of two forces on a body in different directions – the resulting movement may not occur in the direction of any one of the two forces, but in a direction governed by the law of parallelogram of forces. Similarly, the consequences of good and bad actions may not always be experienced as separate, independent good and bad destinies. They may merge creating a unique destiny.

This briefly is the theory of reincarnation and its justification. A fundamental question arises – exactly what reincarnates? Based on the hierarchical structure of a human being, we answer as follows.

What Reincarnates?
Life arises from the contact between causal and subtle bodies, enlivened by *Atman,* and the physical body. When this contact is lost, death occurs. At death:

1. The dead physical body remains.
2. *Atman,* being a part of the infinite, omnipresent *Brahman,* cannot move and remains motionless in the dead body.
3. It is the causal and subtle bodies that transmigrate. Together they determine the conditions of the next incarnation: physical form, environment, tendencies, and challenges.

Bhagavad Gita verse 15.8 explains this thought.

> *When the soul acquires a body*
> *Or leaves it, he carries these*
> *Just as wind carries scents*
> *From place to place.*

The word "these" (defined in the previous verse) means the causal and subtle bodies (see Chapter 5) that transmigrate from the dead physical body to a new physical body, always in contact with *Brahman,* who serves as the medium of transfer.

Process
This transmigration raises several questions.

1. What is the exact mechanism for transmigration? Hindu texts describe what transmigrates but not the mechanics of how causal and subtle bodies move from one location to another. The process is described through analogies – for example, as wind serves to transfer scent.
2. How does the soul choose where to be born? Life begins at conception. The soul takes birth under circumstances that align with its *guna* composition.
3. Are the causal and subtle bodies detectable as they migrate? Hindu philosophy describes causal and subtle bodies as composed of subtle matter that cannot be detected. This creates potential inconsistency: if they are material, they should be theoretically detectable and follow physical laws of movement and interaction. The claim that they are simultaneously material yet beyond sensory perception is resolved by suggesting that subtle body is material but in a non-physical, casual and function sense. Interestingly, science has discovered dark matter, not directly detectable. It shows that the concept of undetectable matter is

not scientifically impossible in the context of present knowledge.
4. Should the memories not be remembered in the present? Hindu philosophy distinguishes between gross memories stored in the brain tissue and tendencies (*vasanas*) stored in the causal body. Specific factual memories decompose at death. However, *vasanas* transmigrate. These create inclinations and aptitudes in the new life without conscious recollection of how they were acquired.

What are carried over from the previous life to the next life are the causal and subtle bodies. The causal body contains unique combination of three *gunas*. A child is conceived with the proportion of three *gunas* present at the child's prior death, endowing each child with unique *vasanas*, or tendencies, at birth. This aligns with our practical experience that children are born with varied tendencies.

Bhagavad Gita verses 14.14 and 14.15 say as much (paraphrased).

If a being dies when Sattva prevails,
He is born among the spiritual.
If Rajas prevails, among those attached to action.
If Tamas prevails, he is born among the deluded.

The verses powerfully suggest that *gunas* at death are carried forward to the next birth, and one is born in circumstances dictated by these *gunas*.

Upon death, the transfer of the causal and subtle bodies from one gross body to another is called reincarnation. Since *Brahman* is omnipresent, the causal and subtle bodies never lose contact with *Atman/Brahman*. Conception occurs when the causal and subtle bodies, enlivened by *Atman,* establish contact with a new physical body, marking the beginning of another cycle of life.

Comparison with Science

Science and reincarnation offer fundamentally different explanations for the same human experiences. Here are key areas where their interpretations diverge:

Life and Death
Scientific view: "From dust we come and to dust we shall return." Life emerges from the organization of matter and ends when biological processes cease. Awareness is produced by brain activity and terminates at death. Death represents the complete cessation of individual existence.
Reincarnation view: Death is merely a transition, like changing clothes or moving from one house to another. Soul is eternal and continues through multiple lifetimes. What we call death is simply the separation of the soul from one physical body before taking on another for new life to emerge.

Personality Differences in Children
Scientific explanation: Children's personalities result from genetic inheritance combined with environmental influences. DNA provides the biological foundation for temperament, intelligence, and behavioral tendencies. Early experiences shape neural development and personality formation.
Reincarnation explanation: Children are born with different personalities because they are old souls carrying tendencies (*vasanas*) developed over countless previous lives. A naturally musical child may have been a musician in past incarnations. Fears, talents, and preferences reflect accumulated experiences from previous births.

Suffering and Inequality
Scientific perspective: Suffering results from natural causes – disease, accidents, genetic disorders, environmental factors, and social circumstances. Inequality stems from historical, economic, and social factors. Much suffering is random and without ultimate purpose or meaning.

Reincarnation perspective: All suffering and circumstances result from one's own past actions through the law of *karma* (see the essay *Karma* Philosophy) combined with systemic *karma*. This provides a moral framework where present inequalities serve a purpose in soul development.

Exceptional Abilities
Scientific view: Exceptional abilities result from favorable genetic combinations, early exposure, intensive practice, and optimal brain development. Some individuals inherit superior cognitive abilities or physical traits that enable extraordinary performance.
Reincarnation view: Child prodigies demonstrate skills, perfected over many lifetimes. A young chess master or musical virtuoso is accessing abilities developed through extensive practice in previous incarnations, carried forward through causal body impressions.

Purpose in Life
Scientific framework: Human life has no predetermined purpose beyond biological survival and reproduction. Progress occurs through random mutations and natural selection, with no ultimate goal or direction. Meaning is created by humans, not inherent in existence.
Reincarnation framework: Life has a clear purpose – the evolution of the soul toward eventual liberation. Each lifetime provides opportunities for spiritual growth. The process is directed by the soul's inherent drive toward happiness and bliss.

These contrasting explanations demonstrate how the same human experiences can be interpreted through entirely different conceptual frameworks, each offering its own internal logic, and explanatory power.

The contrast between scientific and reincarnation explanations raises a crucial question about causation itself. As one example, who determines the circumstances of birth? Examining how each framework explains the birth of a baby illuminates their deeper philosophical differences about agency and responsibility.

Birth of a Baby

On the road from Jaipur to Delhi in India, there was a large billboard on the left-hand side of the road that said, *"It is the child that gives birth to the mother."*

We think that it is the mother who gives birth to the child – with genetic material from both parents. We think that the newborn inherits the tendencies of parents as dictated by genes. The theory of reincarnation is not against genetics. However, it offers a different perspective. It suggests that it is the child who determines the parents, and not the other way around. In this sense, it is the child reincarnating soul who gives birth to the mother.

Science looks at the same event – the birth of the baby – and reverses the cause and effect to say that it is the mother who gives birth to the child. Unlike three thousand years ago, women today have significant control over whether to conceive the child, whether to give birth to the child, the sex of the child, the color of hair and eyes of the child, the health of the unborn baby, and even the mind and intellect of the child.

What is the cause and what is the effect can sometimes be difficult to understand. For example, the faster a windmill rotates, the more wind is observed. This suggests that the rotation of the windmill is the cause and wind is the effect. The reality is exactly the opposite. It is the wind that makes the windmill rotate.

Does the mother give birth to the baby or does the baby give birth to the mother? Rather than debate which explanation is objectively correct, let us ask: *Which perspective encourages moral responsibility among parents?*

- The scientific explanation says that a baby is created by the parents. In a sense, the baby is an invited guest. Therefore, the scientific explanation holds parents personally responsible for the baby that they create.
- On the other hand, reincarnation says that the baby is an old soul attracted to the mother. The baby may be viewed as an

uninvited guest, possibly receiving inadequate care. This may be why Hindu philosophers say – *Atithi Devo Bhava* – *Treat even an uninvited guest as God.*

- A third perspective reframes reincarnation differently. The baby is not simply choosing a mother, but rather arriving as an old soul on its path to liberation. In this view, parents function as custodians or trustees, entrusted with guiding the child on its spiritual journey. The soul, recognizing them as the best possible guardians, places itself under their care. This interpretation shifts parental responsibility from mere biological duty to a sacred trust – a moral obligation to nurture the child's onward progress with wisdom and compassion.

Science fosters morality by holding parents personally responsible for the children they willingly choose to have and believe belong to them. Reincarnation, in contrast, cultivates moral responsibility by framing parenthood as a divine stewardship of the child's soul. Religion has long served as a foundation for ethical guidance. However, if scientific explanations were to lead to stronger moral accountability, then the role of religion may gradually evolve – or even diminish.

Beyond these fundamental philosophical differences, reincarnation and scientific thinking share a surprising structural similarity. Both describe evolutionary processes that follow remarkably parallel mechanisms, though operating in different domains.

Darwin Vs. Reincarnation

Darwinian evolution is the evolution of physical body. It says that the evolution of gross body, mind, intellect, and awareness are a result of the evolving physical body. To compare the theory of reincarnation and the theory of evolution, we first outline the basic mechanism of Darwinian evolution in a simplified manner.

Darwinian Evolution

Let us suppose that on a remote island, there is a species of birds, some with short beaks and others with long beaks. Such physical variations exist within a population. Birds with long beaks tend to produce descendants with long beaks, but not always. The same is true for birds with short beaks.

Now, suppose environmental conditions change so that food becomes harder to access, favoring birds with longer beaks. Nature introduces a challenge, and only birds with longer beaks can efficiently reach food sources. As a result, they survive and reproduce at higher rates, while birds with shorter beaks struggle to feed and often fail to reach reproductive age. Over several generations, short-beaked birds die out, leaving only long-beaked individuals. The population, *through natural selection*, adapts to its environment.

On another island, where different environmental conditions exist, evolutionary adaptations would take a different course. This gradual and slow process requires extended periods for significant physical changes to manifest. This is Darwinian evolution in its simplest form.

The basic mechanism of evolution of physical body may be summarized as follows.

1. **Goal**: Biological evolution does not have a predetermined goal. It is an ongoing, natural process driven by genetic variation and environmental pressures. While evolution produces increasingly complex and well-adapted organisms, it does not *aim* for a specific endpoint. It was not the goal of Darwinian evolution to produce human beings. Species evolve in response to their environments, meaning that evolution is dynamic, not purposeful.
2. **Variation**: Within every species, individuals exhibit differences in traits such as size, shape, color, and strength. These variations arise through genetic mutations and recombination, ensuring diversity within the population.

3. **Inheritance**: Traits are passed from parents to offspring through genetic material. While descendants of a particular body form often inherit similar characteristics, variation persists across generations.
4. **Challenge**: As environmental conditions change, nature presents challenges that affect survival. Certain body forms may be better adapted to cope with these challenges than others.
5. **Selection**: Organisms with beneficial traits have a higher likelihood of survival and reproduction. Those with disadvantages may struggle to thrive and fail to pass on their genes. Over time, natural selection favors those most suited to the environment. This process is often called *survival of the fittest*.
6. **Adaptation**: Populations gradually adjust to their surroundings. Traits that enhance survival and reproduction become more widespread, shaping the species in response to environmental pressures.
7. **Time**: Evolution occurs over many generations, accumulating small changes that eventually lead to significant transformations in species. This process is slow but persistent, shaping the diversity of life across eras.

Reincarnation

Hinduism's theory of reincarnation says that it is the soul that reincarnates and evolves until it becomes one with *Atman*. Soul is conditioned consciousness, namely, consciousness limited by the causal and subtle bodies. The concept of conditioning may be understood as follows: Suppose that I can drive a car and fly a plane. If the only body available to me is a car, I cannot demonstrate my skills at flying the plane. Similarly, consciousness gets limited by the causal and subtle bodies, and its perceptions and actions are limited by the gross body it occupies.

The proposed mechanism of reincarnation or evolution of conditioned consciousness is as follows:

1. **Goal:** Reincarnation may be thought of as a directed process – directed not by external forces arbitrarily guiding souls, but by

the inherent arrow of happiness within everyone. The argument goes as follows. Happiness is an important goal in life. Given life's challenges, one acts to be happy and avoid suffering. The behavioral patterns so developed become ingrained tendencies and shape future births. Over time, these tendencies refine one's ability to find deeper and more sustainable happiness, culminating in bliss. *This interpretation makes reincarnation a self-directed, meaningful, and profoundly optimistic process, ensuring eventual liberation of all, and not a reward and punishment process.*

2. **Variation:** There is variation in conditioned consciousness among living beings because we all have different causal, subtle, and gross bodies. Everyone's tendencies and circumstances in life are different. Past actions shape these differences.
3. **Inheritance:** The causal and subtle bodies at death reincarnate as causal and subtle bodies at the next conception.
4. **Challenge:** Everyone is presented with unique challenges in the form of destiny based on their prior un-fructified actions that are to fructify in this life, called *prarabdha karma*.
5. **Selection:** Some individuals are better able to cope with their destiny than others, by exerting appropriate self-effort.
6. **Adaptation:** The evolution of the soul depends upon how one responds to destiny in each life, ultimately reaching the stage of being liberated from suffering. Those who respond wisely accelerate their spiritual evolution, while others may remain entangled in repeated patterns.
7. **Time:** The changes usually occur extremely slowly, requiring thousands or millions of lifetimes to reach destination.

There are many parallels and some differences between the theories of evolution and reincarnation. Both describe a process of gradual transformation. One involves biological transformation of the physical body and the other involves transformation of causal and subtle bodies. Both have a mechanism for change – evolution operates through random genetic mutations and natural selection, whereas reincarnation is driven by the law of *karma*, where past actions influence future births.

The differences are that reincarnation presents a *directed* evolution of conditioned consciousness, while Darwinian evolution describes an *undirected* evolution of species through environmental pressures. The mechanism of evolution proposed by the theory of reincarnation is broader than that proposed by Darwin, as it allows for uniquely different individual challenges. It may also be better able to cope with scientific intervention in human genetics than the mechanism of Darwinian evolution.

Finally, evolution of conditioned consciousness must mean the evolution of the conditioning factors, namely, causal, and subtle bodies, because consciousness itself does not need to evolve. Consciousness is *Brahman*. And evolution of causal and subtle bodies may necessitate evolution of physical body, bringing the two theories conceptually closer together.

Benefits of Reincarnation Assumption

Reincarnation is a hopeful thought. It offers several benefits – both practical and philosophical. Here are some key advantages:

- Reincarnation allows individuals to refine their souls over multiple lifetimes. Each life offers new experiences, challenges, and lessons, helping the soul evolve toward self-realization. The practical implication is that mistakes are not final; they become steppingstones for improvement.
- Reincarnation provides a mechanism for moral justice – good deeds not rewarded bring favorable circumstances in future lives, while harmful actions create challenges that must be resolved. This process gives meaning to life's ups and downs.
- Reincarnation reduces the fear of death by making it a transition rather than an end. One can view life as a continuous

- progression, knowing that whatever is left unfinished in one lifetime can be pursued in the next.
- Reincarnation makes it easier to accept ourselves and our destiny – our personality, faults, tragedies, and suffering in our life. It also makes it easier to accept others.
- Reincarnation allows us to view parenting as stewardship – trusteeship of a soul previously born many times.
- The concept of past-life tendencies being reborn suggests that individuals carry their strengths, talents, and inclinations forward. This explains innate abilities, and an intuitive sense of purpose that some people feel early in life.
- If each life is driven by the pursuit of happiness – minimizing suffering and seeking fulfillment – then reincarnation can be seen as a self-guided process toward bliss. Over many lifetimes, individuals refine their ability to reduce suffering, eventually leading to liberation from suffering.

Do you want to come back again?

Do you seek an end to the cycle of birth and rebirth, or would you choose to return once more?

At a practical level, this question may seem irrelevant, as we do not appear to retain memories from past incarnations – though anecdotal claims suggest otherwise. Unlike waking up after sleep, where past experiences shape future actions, reincarnation does not seem to carry direct, conscious recollection from one life to the next.

Moksha is often defined as liberation from the cycle of birth and rebirth. However, another perspective views it as freedom from suffering within a lifetime. Regardless of one's belief in reincarnation, few would claim to have lived so impeccably as to merit complete escape from rebirth.

In my own case, if reincarnation is real, I have made enough mistakes to say that I would welcome another opportunity at life. Those truly deserving of *moksha* might also choose to return – not for their own sake, but to guide others toward liberation. We are all likely to be on this journey together!

Practical Applications

Reincarnation assumption has several practical advantages to enhance daily life and personal development. The concept of viewing mistakes as learning opportunities rather than permanent failures can foster resilience and growth mindset, encouraging people to persist through challenges with the understanding that setbacks contribute to long-term development. The *karmic* perspective on suffering and inequality can help individuals accept difficult circumstances with greater equanimity while taking responsibility for their actions, knowing that both challenges and blessings serve a purpose in personal evolution.

The essay's interpretation of parenting as stewardship rather than ownership can transform family dynamics, encouraging parents to view their role as nurturing guides who support their children's unique paths rather than imposing their own expectations. This perspective can reduce parental anxiety and control while fostering deeper respect for a child's individual nature and capability. Similarly, understanding that people carry forward tendencies and talents from their past experiences can help educators and mentors recognize and cultivate innate abilities rather than forcing uniform development.

The reincarnation framework's focus on spiritual evolution through the pursuit of genuine happiness or bliss can guide decision-making toward choices that promote long-term fulfillment rather than temporary pleasures. This can encourage individuals to prioritize character development, meaningful relationships, and service to others over material accumulation. The concept of multiple lifetimes for growth can also reduce the pressure to achieve everything in one lifetime, promoting patience with personal development while maintaining motivation for continuous improvement.

Finally, the essay's comparison between biological and spiritual evolution suggests that personal growth follows natural laws of gradual adaptation through challenges. This insight can help people embrace life's difficulties as necessary catalysts for development, approach change with patience, and trust the transformative power of sustained effort over time, creating a more compassionate and purposeful approach to living regardless of one's metaphysical beliefs.

Summary

This essay examines Hindu philosophy's theory of reincarnation, which holds that upon death, the Soul (causal and subtle bodies in contact with *Atman*) transmigrates to a new physical body in an endless cycle until achieving liberation. The *Bhagavad Gita* presents this concept when Krishna tells Arjuna that souls are eternal, simply changing bodies like clothing.

Reincarnation is necessitated by past actions who effects are yet to be experienced. What transmigrates is soul – casual and subtle bodies enlivened by *Atman*. Rebirth continues until the soul achieves union with *Brahman*.

The essay contrasts this with scientific materialism. Where science explains personality through genetics and environment, reincarnation attributes it to past-life tendencies. Science views suffering as random; reincarnation sees it as *karmic* lessons. Regarding birth, science emphasizes parental choice and genetic inheritance, while reincarnation suggests souls choose their parents and circumstances.

Interestingly, both Darwinian evolution and reincarnation share structural similarities – gradual transformation through challenges over long periods. However, biological evolution is undirected and environmentally driven, while spiritual evolution is self-directed by the soul's quest for liberation and *karmically* driven.

Reincarnation offers practical benefits: hope for improvements through multiple lifetimes, moral justice explaining life's inequalities, reduced fear of death as mere transition, acceptance of oneself and circumstances, and viewing parenthood as spiritual stewardship. *The theory provides meaning by framing each life as an opportunity for growth toward eventual freedom from suffering, with the optimistic view that all souls will eventually achieve liberation.*

CHAPTER 7
KARMA PHILOSOPHY

Have you ever wondered: *Why do bad things happen to good people? Is there any justice in this universe – cosmic or otherwise? Do my actions even matter in the grand scheme of things?*

These timeless questions have stirred minds for centuries. The karma philosophy of Hinduism steps in to explain suffering, agency, and cosmic order through the lens of causality and reincarnation.

Building on earlier discussions, this essay articulates the governing laws of karma, destiny, and rebirth, and introduces systemic karma – a broader, more holistic framework that includes not just individual actions but also universal actions.

We will explore how causal body, gunas, vasanas, types of karma, destiny, action, and free will interweave to form karma theory. This is not just a doctrine of reward and punishment but is a theory of evolution, directed by the "arrow of happiness" that shapes the evolving trajectory of the soul toward liberation.

Human beings possess an innate urge to seek reasons for the events that unfold around them – especially when those events are personal and painful. Questions like "Why me?" arise: Why must someone who has led a virtuous life undergo a failed major surgery? Why is a child born with congenital defects? Why are some born into wealth while others struggle for basic sustenance? The underlying causes of such disparities are seldom obvious or emotionally acceptable. In such moments, many turn to *karma* philosophy for both explanation and solace.

Karma – meaning action – links present circumstances to past choices across lifetimes. Our current destiny is shaped by actions from both this and prior lives, while our future will be sculpted by how we respond to the present. Regardless of personal belief in this worldview, the goal of this essay is to offer a logically coherent and accessible exposition of *karma* philosophy, grounded in our shared practical understanding of causality as presented in Chapter 3.

To that end, the discussion begins by summarizing key insights from our earlier essay, *Cause and Effect*, which are recontextualized here to support the *karma* framework:

1. Destiny results from past actions.
2. Actions yield both external outcomes and internal impressions.
3. Causes and their effects may be separated by vast stretches of time and space.
4. The principle "as you sow, so you reap" holds in many – but not all – cases.
5. Outcomes often hinge on surrounding conditions.
6. Some causes lie within our control; others do not.

Karma philosophy emerges naturally from these observations when we include our prior assumption that the soul reincarnates and causality unfolds across multiple lifetimes, allowing unresolved effects to manifest in the future.

Karma Theory

Karma philosophy says that our present destiny is shaped by our past actions – whether from this life or previous lives – and our current actions help create our future. While actions in this life often account for our present experiences, they cannot explain circumstances such as the conditions of our birth, unexpected tragedies, or the timing and nature of death. Rather than attributing such events to luck or divine will, Hindu philosophers, deeply rooted in their belief in causality and personal accountability, developed a comprehensive framework that integrates reincarnation to explain life's disparities.

At its core, *karma* philosophy rests on three interwoven laws.

The **Law of Destiny** asserts that what we face in this life results from our own past actions, not from chance or divine predestination. Destiny unfolds through cause and effect, making us the architects of our own lives – there is no one else to blame.

The **Law of *Karma*** (Action) emphasizes that our response to present circumstances determines our future. While our *vasanas*, or innate tendencies, often guide our behavior, human beings possess free will that enables them to act consciously. Although we cannot alter the destiny we face today, we can shape tomorrow's outcomes through deliberate and wise action.

The **Law of Reincarnation** extends the reach of causality across lifetimes: every action must yield its effect, and if unresolved within a single life, rebirth becomes necessary. Thus, the destiny we encounter at birth – including the life conditions of a child – is shaped by past actions whose consequences have yet to manifest.

Yet, these laws can be misapplied if taken too literally. To assume that a child born into suffering committed grave wrongs in a past life – or that one born into privilege must have lived virtuously – oversimplifies a profoundly nuanced philosophy. While the principle "as you sow, so you reap" may often hold, causality is complex. The effects of action are influenced by a host of factors, including social dynamics, environmental context, and interdependent events beyond an individual's control. For this and other reasons, reducing life circumstances to moral judgment is not only inaccurate but also antithetical to the spirit of *karma* philosophy, which encourages understanding over blame.

Systemic Karma

Our destiny and our responses to it are shaped not only by our own actions but also by the actions of other entities. As individuals, we are part of an intricate system that includes family, friends, society, nature, planet, and the universe. The concept of *systemic karma* refers to the aggregate actions taken by these other entities, and it exerts significant influence on our experiences, choices, and even ingrained tendencies.

Traditional *karma* philosophy often emphasizes personal responsibility – what we do leads to what happens to us, and how we respond shapes our future. While this focus is important, it tends to overlook the extent to which external forces condition both our circumstances and our capacity to act. A helpful analogy comes from astrophysics: although dark matter and dark energy constitute 95% of the universe, science focuses largely on the 5% of matter we can observe and influence. Similarly, conventional interpretation of *karma* philosophy highlights direct personal action, even though *systemic karma* plays a vital role.

Systemic karma explains why many life experiences lie beyond individual causality. The effects of inventions like the automobile, economic events like the Great Depression, or tragedies like the Holocaust ripple through history, affecting individuals who had no role in their creation. Advances in medicine, the eradication of smallpox, or natural disasters such as tsunamis exemplify how non-personal events can significantly alter human destiny. Our lives are embedded within vast networks: family, society, nation, planet, and cosmos.

Systemic karma also shapes the menu of options available to us. Whether traveling between cities or seeking medical care, the range of choices we enjoy exists because of systemic progress, not solely personal merit. The improvements we benefit from – longer life expectancy, better technology – are not the *karmic* fruits of our own making, but of a collective inheritance. *Systemic karma* is a double edged sword.

Consider this illustrative parable: A lady instructs her driver to take her to the temple in her car – an Aston Martin. In a hurry, she urges him to speed through a shortcut, only for the car to end up in a ditch. The tire, suddenly lodged in muddy earth, wonders, "Why am I stuck here?" A simplistic *karma* explanation might say the tire had a past life affinity for mud – an echo of swinish indulgence now coming full circle. But this neglects the larger system at play. The tire's fate was not simply its own doing – it was entangled with the decisions of the lady, the driver, the capabilities of the car, and the systemic neglect of the road. Their destinies were interlinked.

This parable highlights how suffering or success cannot always be ascribed to individual actions alone. Interpretations of karma philosophy often ignore this systemic dimension, leading to misguided moral judgments. Without acknowledging systemic karma, we risk blaming victims or congratulating beneficiaries without a full understanding of causality.

Recognizing *systemic karma* has several implications.

First, under the **Law of Destiny**, we are reminded that suffering is not always deserved – external forces often contribute, and it is unjust (and incorrect) to assign blame blindly.

Second, under the **Law of *Karma***, accepting systemic influence does not excuse inaction. Instead, it reinforces our responsibility to act wisely and to seek help when needed. Present choices still matter deeply.

Lastly, under the **Law of Reincarnation**, *systemic karma* opens the possibility that liberation may not be purely individual. Just as the lady, driver, and car must reach the temple together, spiritually awakened beings may choose to reincarnate in service of collective upliftment.

By integrating *systemic karma* into the framework, *karma* philosophy becomes more nuanced, compassionate, and aligned with lived reality. It recognizes that destiny is co-created – a result of personal actions coupled with *systemic karma*. Personal actions determine how we respond, while *systemic karma* defines the arena in which that response occurs. Together, they form a more holistic and realistic vision of human experience – one that honors individual agency without overlooking the profound interconnectedness of all life. A refined understanding of *karma* theory recognizes both individual effort and collective influence, making it a holistic process rather than a rigid system of personal reward and punishment.

Functional Relationships

The functional relationships of *karma* theory describe the relationships between various elements including different types of *karma*, *gunas*, causal body, *vasanas*, destiny, free will, and actions. Understanding these relationships is essential for comprehending how our actions influence our present and future lives. We describe these relationships in words, graphically, and in equation form as conceptual models, not calculable formulas. We begin with definitions.

Karma: *Karma* means action. In the context of this theory, *karma* includes intentional actions – mental, verbal, or physical – freely taken by human beings, involving enjoyment or suffering. Unintended actions are not counted as *karma*. Actions taken by animals, and young children are not counted as *karma* because of a lack of intent.

Intent matters is a well-recognized principle in U. S. jurisprudence. The consequences for an individual are dramatically different if he commits murder, kills someone in self-defense, or kills an enemy soldier in a just war even though the result is the same – somebody is dead.

Our actions produce effects, and we must experience the results of these actions. Some consequences are experienced in this life, while others may be delayed, and experienced in later lives. Therefore, Hinduism proposes several categories of *karma:*

1. ***Sanchit karma***: This encompasses all prior actions, good or bad, accumulated over past lives whose effects have yet to be experienced by us. Good actions do not cancel bad actions, actions simply accumulate.
2. ***Prarabdha karma***: It is our practical experience that the time lag between an action and its effect is variable and indeterminate. *Prarabdha karma* is that portion of our past actions (*sanchita karma*) whose effects we are to experience in the

present life. Good actions may produce good effects, and bad actions may produce bad effects, but it is our practical experience that joint effects of multiple actions can be quite unexpected, leading to complex life experiences.

Note that *sanchit karma* usually consists of many good deeds and many bad deeds, because it includes un-fructified deeds from innumerable past lives. *Prarabdha karma* is only a small portion of the *sanchit karma*. It could happen that, by chance, only good deeds come to fruition in this life making a person feel that he must have led exemplary past lives to deserve such good fortune – not realizing that the next life may bring a string of bad deeds to fruition. The reverse is also true. The conclusions we draw from what meets the eye can be quite wrong.

3. ***Agami karma***: These are our thoughts and actions in this life which are under our control. Present actions influence the future allowing us to shape our destiny. For instance, getting vaccinated can prevent severe illness, modifying the effect of *prarabdha karma*. Those actions whose effect is not experienced in this life are added to *sanchit karma*.

4. ***Systemic karma:*** These are actions taken by entites other than ourselves.

Gunas: (See Chapter 5). These are three fundamental qualities of *Prakriti* (primordial matter) known as *sattva, rajas*, and *tamas. Gunas* are un-manifest. *Sattva* manifests as thoughts in a state of equanimity. *Rajas* manifests as passionate, desirous, and agitated thoughts. *Tamas* manifests as inertia, lethargy, and lack of intellectual conviction.

Causal body: (See Chapter 5). Everyone has a unique proportion of the three *gunas* at any given time. This combination of the three *gunas* called the causal body of that individual at that time. It determines our tendencies (*vasanas*).

Vasanas: (See Chapter 5). These are our tendencies determined by our unique causal body. *Vasanas* have considerable, but not total, influence on our body, mind, and intellect. *Vasanas* are un-manifest and manifest themselves through thoughts and actions. Through proper control of our mind and actions, our inborn *vasanas* can be altered. Present *vasanas* set the pattern; our self-effort can alter them, by changing the makeup of our causal body.

Destiny: What we face in life is destiny. The circumstances at the time of our birth are destiny. Our destiny at any time during our present life is a result of our *prarabdha karma, agami karma,* and *systemic karma* and does include the results of unintended actions. Although we cannot directly change our present destiny, self-effort allows us to reshape our future through conscious actions.

Free will: Human beings are endowed with free will, freedom to generate alternatives and choose from them. Free will is a part of our intellect. Among living beings, humans may be the only ones with significant free will. Other living beings act primarily on instinct. Our free will is limited. It operates within constraints – we can choose from available options, but the range of options is determined by other factors. Free will allows us to exert self-effort which can act independently of *vasanas*, shaping our future.

Actions: We act in response to our destiny. Our actions are governed by what we face (destiny), our tendencies (*vasanas*), our free will, and systemic actions. By understanding and managing these elements, we can navigate our lives more consciously.

These factors work together. Consider planning a party where you are responsible for food (destiny). Your natural preference might be to prepare Indian cuisine (*vasanas*), while friends offer to bring additional dishes (*systemic karma*). However, learning that Japanese guests will attend, you consciously decide to prepare some Japanese food despite your preferences (free will). This demonstrates how free will operates as conscious choice-making within the context established by destiny, *vasanas*, and systemic influences. The actual performance of actions is constrained by our capabilities.

Understanding the functional relationships in *karma* theory provides valuable insights into how our actions influence our present and future lives. By recognizing the interplay between *karma, gunas*, causal body, *vasanas*, destiny, free will, and actions, we can make more informed choices and lead a balanced life.

Figure 7.1 graphically illustrates the relationships between these terms. Note that *Atman* stays above the fray and does not participate in taking actions or in their fruits. Numbers 1 to 6 in Figure 7.1 relate to the six functional relationships that follow.

Figure 7.1: *Karmic* relationships

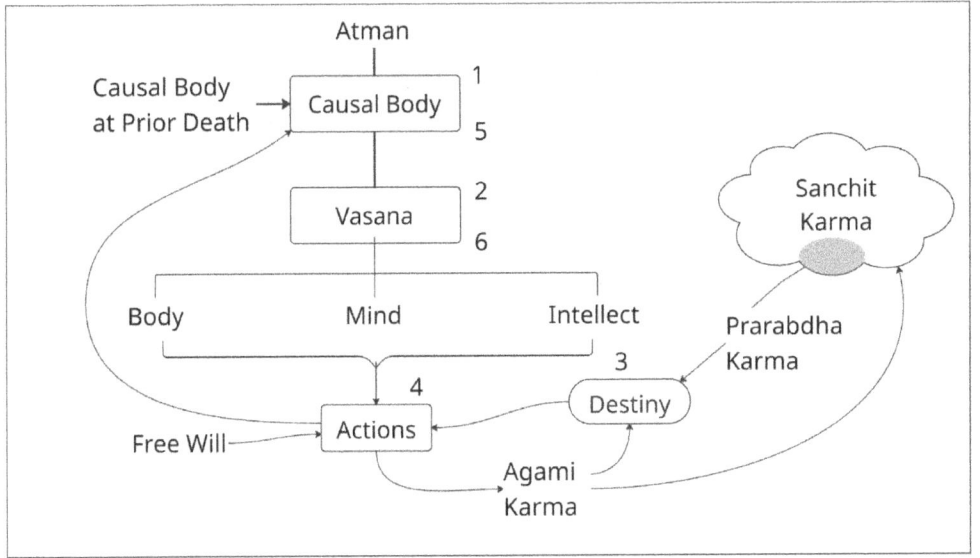

It should be noted that karma resides in the causal and subtle bodies. Similarly, free will resides in the intellect. They are shown separately in Figure 7.1 for ease of display. Also, actions, influenced by vasana, are subject to the limitations of our body.

For those inclined to examine *karma* theory through a mathematical lens, we present the relationships between key components in equation form – not for numerical computation, but for conceptual clarity (**others can skip this section**). To simplify the framework, we treat *systemic karma* as a separate influence. Those who prefer to subsume it under *prarabdha karma* may simply omit the additional terms denoted by ε (epsilon).

1. *Causal body (conception) = Causal body (prior death)*

   It is the causal and subtle bodies that transmigrate across lifetimes (see Chapter 6), perpetually in contact with *Brahman*. In reincarnation, what carries forward is the specific proportion of the three *gunas – sattva, rajas,* and *tamas –* in the causal body. This ratio remains unchanged from prior death to new conception, though it may be altered during gestation due to systemic influences. The three *gunas* always add to 1 or 100%. Their respective fractions at birth may be denoted by $\alpha_0, \beta_0,$ and γ_0.

   The causal body at birth may be written as

   Causal body (birth) $= \alpha_0$ *Sattva* $+ \beta_0$ *Rajas* $+ \gamma_0$ *Tamas*
   Where $\alpha_0 + \beta_0 + \gamma_0 = 1$

   If $\alpha_0, \beta_0,$ and γ_0 are 50%, 30% and 20% respectively, then the causal body at birth is 50% *sattvic*, 30% *rajasic*, and 20% *tamasic*.

2. *Vasanas (birth)* $= f_0(\alpha_0$ *Sattva* $+ \beta_0$ *Rajas* $+ \gamma_0$ *Tamas*)   Where $\alpha_0 + \beta_0 + \gamma_0 = 1$

   Vasanas (tendencies) are a function of the proportion of the three *gunas* as shown by the above equation. *Gunas* at birth are like the *gunas* at prior death if we neglect the effect of pregnancy. Then, *vasanas* at birth are like the *vasanas* at prior death ensuring continuity of causal body from prior death to new conception and birth.

3. *Destiny at time t* $= f_1$ *(prarabdha karma, and agami karma that has produced effects by time t)* $+ \varepsilon_1$

   Destiny at any point in life is shaped by *prarabdha karma* and the effects of *agami karma* up to that point. The ε_1 term represents *systemic karma –* external factors such as societal conditions or natural events that also shape our destiny. For instance, a child born with AIDS in sub-Saharan Africa is affected not only by *personal karma* but also by the *systemic karma* of family and society. If a cure later eliminates this condition for future generations, it reflects the *systemic karma* of those who advanced the cure.

4. *Actions at time t = f_2 (destiny at t, vasanas at t-1, free will) + ε_2*

    Our conscious actions are responses to destiny, influenced by prior *vasanas* and the active role of free will. When *vasanas* align with virtuous conduct, action becomes easier. When they do not, free will must work harder to guide us toward right behavior. The term ε_2 accounts for systemic contributions to our actions – support or obstacles created by others or by the environment. Actions are subject to limitations of our body.

5. *Causal body* at time t = α_t Sattva + β_tRajas + γ_tTamas Where $\alpha_t + \beta_t + \gamma_t = 1$
   Gunas at time t = f_3 (*Gunas* at birth, actions from birth till time t-1) + ε_3

    Gunas that constitute causal body at time t are respectively denoted by α_t, β_t and γ_t. *Gunas* evolve through life, influenced by the original composition at birth and modified by one's actions. *Systemic karma* (ε_3) also plays a role in shaping *gunas* over time.

6. *Vasanas at time t* = $f_4(\alpha_t$ Sattva + β_tRajas + γ_tTamas) Where $\alpha_t + \beta_t + \gamma_t = 1$

    Our current *vasanas* are a direct outcome of our present causal body. Thus, behavioral tendencies are dynamic, shaped by past impressions, ongoing effort, and external influences.

Why Do Bad Things Happen to Good People?

We now turn to an interesting philosophical and practical question, "Why do bad things happen to good people?" The question could also be framed as "Why do good things happen to bad people?" or more generally, "Why do things happen to people?" The fact that "things happen" has been expressed more colorfully elsewhere. Things happen to all people. Sometimes it is hard to define what a "bad thing" is and who the "good people" are. An only child works hard, grows up to be a doctor, flies off to a medical conference in Europe, the plane crashes, and he dies. His parents wonder – was his becoming a doctor a good thing or a bad thing? He would be alive today had he become an accountant. The terms 'good people' and 'bad things' are inherently subjective, culturally relative, and often apparent only in retrospect. We use them here as they reflect the common way people frame these questions when turning to *karma* philosophy for understanding.

At a practical level, we usually feel that we know what a bad thing is and who the good people are. From the *karma theory* equations stated above, our destiny is a function of actions in previous lives (*prarabdha karma*), actions in this life (*agami karma*), and systemic actions (*systemic karma*). When bad things happen, sometimes we understand their reasons, and sometimes we do not. We may know that what happened is because we did something wrong – an auto accident caused by excessive drinking (*agami karma*). We may know that what happened is due to the system we are a part of – a tornado that destroys our house (*systemic karma*). When we do not know the cause, we are left with at least three options:

1. To say that science will be able to find the cause someday. This is looking at the world through deterministic materialism. One example is smallpox – a deadly disease whose cause was unknown leading to innumerable fatalities till science discovered the cause and eradicated the disease.
2. If the laws of nature are inherently non-deterministic, as increasingly appears to be the case, then one may have to accept probabilistic emergence – chance or luck as a reason. For example, John and David are deployed in a war, John is killed but David returns safely. If the laws are inherently probabilistic, science will not be able to predict with certainty: *"John was always going to die, and Peter was always going to live."* One would have to accept chance or luck as an explanation.

3. *Karma theory* provides another explanation based on causality and continuity of life beyond death. This is *prarabdha karma,* not known but never accidental. It provides a structured moral causality to what is otherwise inexplicable, thereby, potentially helping answer the "why me?" question.

Hinduism believes in reincarnation, a process in which causal and subtle bodies take repeated births and evolve such that a person eventually becomes *Jivanmukta* – liberated while living, free of suffering. This evolution occurs because of self-effort in response to our destiny. Why do bad things happen to good people? As stated earlier, our destiny at any time depends upon three things: what we have done in our previous lives, what we have done in this life, and the systemic effects we experience. Regarding the effects of what we have done in this life, science can sometimes provide exact answers; at other times it can only provide a statistical answer in terms of the probability of that event happening, and at other times may not have any answer. Effects caused by systemic causes may sometimes be understandable, at other times not. Our actions in previous lives are unknown. When there is no clear explanation, Hinduism's answer to the question "why do bad things happen to good people?" is – *Because of our prior actions in this or previous lives whose effects are to fructify in this life, coupled with the effects of systemic karma.*

If the question were "***when*** bad things happen to good people?" – the answer would be – *Focus on your self-effort because how we respond to the things that happen helps decide the course of evolution of our soul and our ultimate liberation. Self-efforts must be directed toward reducing suffering and achieving happiness. In life after life, the soul learns to be happy, reaching a stage of bliss. This arrow of self-directed happiness makes us Jivanmukta.*

Practical Applications

Understanding *karma* philosophy offers several practical benefits for navigating life's challenges and making conscious choices. In decision-making, we can apply a three-factor analysis before important choices: examining our current destiny or situation that we must work with (*agami, prarabdha* and *systemic karma*), recognizing what our natural tendencies are pushing us toward (*vasanas*), and determining what conscious choice suggests is wisest (free will). For instance, when choosing a career while family expects one path, natural inclinations point toward another, and conscious reflection suggests a third option that serves both personal growth and others' needs, this framework provides clarity for navigating competing influences.

When facing suffering or unexpected challenges, *karma* philosophy helps reframe the question from "Why me?" to "How now?" This shift moves us from seeking blame – whether individual or systemic – toward asking how to respond consciously. While maintaining compassion for genuine systemic injustices, this approach reduces victim mentality and empowers constructive action. The framework helps distinguish between aspects of difficulty that might stem from personal actions requiring inner work, systemic issues requiring collective action or acceptance, and mixed situations requiring both individual and social responses.

In relationships and social interactions, understanding *systemic karma* prevents the misapplication of *karmic* blame – avoiding the assumption that someone's suffering necessarily reflects their moral failings. Instead, we can respond with greater compassion while still maintaining appropriate personal responsibility. Since how we respond to circumstances shapes future *karma*, developing practices to act with awareness rather than simply reacting becomes essential for spiritual evolution.

For personal development, the philosophy encourages *guna* awareness – noticing when we act from *sattva* (clarity), *rajas* (agitation), or *tamas* (inertia) – and consciously choosing actions that cultivate more *sattvic* responses over time. Recognizing that our causal body evolves through actions, we can work to modify ingrained tendencies (*vasanas*) through persistent conscious choice, understanding that change is possible even when challenging.

Most importantly, *karma* philosophy provides mental satisfaction and meaning when no other explanation exists, making acceptance of difficult circumstances psychologically possible. Rather than being trapped in "random universe" despair or chronic resentment, the framework offers a coherent narrative where events have purpose, even if unknowable in detail. This does not promote passivity but enables people to accept their current circumstances as a meaningful starting point for conscious action.

Finally, the *systemic karma* framework reminds us to seek help from others when responding to challenging situations. Since systemic influences significantly shape our circumstances and available options, consciously drawing on the knowledge, resources, and support of others becomes a wise application of the philosophy rather than isolated individual struggle. Even when facing results of our own past actions, we can respond more effectively through collective wisdom and support. Seeking help is itself a conscious choice. Just as negative systemic influences can affect us, positive systemic resources – mentors, communities, institutions, and collective knowledge – are available to support our conscious evolution toward greater happiness and eventual liberation.

Summary

This essay presents *karma* philosophy as a sophisticated system for understanding life's inequities, moving beyond simplistic reward-and-punishment interpretations to embrace a nuanced view of causality across lifetimes. The philosophy rests on three fundamental laws: the Law of Destiny (present circumstances result from past actions), the Law of *Karma* (current responses shape future outcomes), and the Law of Reincarnation (causality extends across lifetimes). A key addition is *systemic karma* – the aggregate influence of collective actions from society, nature, and the universe that profoundly shapes our experiences, explaining why individual actions alone cannot account for all life circumstances.

The philosophy operates through three types of *karma: Sanchit* (accumulated from past lives), *Prarabdha* (experiencing effects now), and *Agami* (current actions shaping the future), coupled with *systemic karma*. These work alongside the three *gunas* that form our causal body and determine our tendencies (*vasanas*), while free will provides our capacity to make conscious choices despite these influences.

To "Why do bad things happen to good people?" the essay offers a compassionate response: destiny results from the complex interplay of *agami karma*, *prarabdha karma* and *systemic karma*. Rather than assigning blame, the focus should be on conscious response. Self-directed efforts toward reducing suffering and achieving happiness create an evolutionary "arrow of happiness" that guides the soul through lifetimes toward ultimate liberation and bliss.

In practical application, this *karma* philosophy provides valuable guidance for conscious living. It offers a three-factor framework for decision-making that considers destiny, *vasanas*, and free will, while encouraging the shift from "Why me?" to "How now?" when facing difficulties. The philosophy prevents harmful *karmic* blame-casting and promotes compassionate responses to suffering. It supports personal development through *guna* awareness and conscious modification of ingrained tendencies. Most significantly, *karma* philosophy provides mental satisfaction and meaning when no other explanations exist, making psychological acceptance of difficult circumstances possible without promoting passivity. The framework also encourages seeking help from others as a wise application of *systemic karma*, recognizing that conscious evolution benefits from collective wisdom and support rather than isolated individual struggle.

CHAPTER 8
MATHEMATICS OF CHATUR VARNA

What if social classes were not rigid hierarchies, but reflections of your innate qualities – and tools for living a meaningful life?

Chatur Varna means four social classes. This essay offers a fresh, analytical look at the ancient varna system of Hinduism, reinterpreting it as a framework for understanding human tendencies, contributions, and spiritual growth.

Building on the Sankhya philosophy of Prakriti and gunas – sattva (balance), rajas (activity), and tamas (inertia) – the essay reveals how the original fourfold classification (Brahmin, Kshatriya, Vaishya, Shudra) may have emerged from patterns in human behavior and potential. With the help of a mathematical model, it tackles deep questions: Why exactly four varnas? How are they formed? Can you change your varna – or transcend it altogether?

Whether you are curious about the philosophy of personality, the ethics of social order, or the mathematics of human nature, this journey offers insights into how ancient wisdom can guide modern self-discovery.

Over time, the original *varna* system of Hinduism was misapplied and morphed into what is now known as the caste system. Unlike the *varna* system, which classified individuals based on their qualities and actions, the caste system assigns a person's status by birth, often leading to discrimination and social stratification. This caste system has been a source of extreme discrimination in India. Since India's independence, the caste system has been outlawed, yet its deep-rooted presence has made societal change slow and arduous.

However, many factors are beginning to reduce the role of the caste system, both in the workplace, and in society. One significant factor is the increasing participation of women in the workforce. As a result, arranged marriages are gradually being replaced by "love" marriages, where caste plays a less important role.

The purpose of this essay is not to discuss the caste system but to explore the original philosophical *varna* system of Hinduism, its underlying logic, and its universal applicability even today.

Understanding the distinction between *varna* and caste is crucial for appreciating the original intent and the modern implications of this ancient classification. By revisiting the principles of the *varna* system, we can strive for a society where individuals are valued for their qualities and actions rather than their birth.

Bhagavad Gita

In Chapter 1 of the *Bhagavad Gita*, Arjuna articulates a series of misconceptions, which Krishna later identifies and addresses. These misconceptions, common even today, include the notions that death occurs solely when the physical body dies, that attachment can override duty, and the belief that an individual's *varna*, or social class, is determined by birth.

Throughout the remaining seventeen chapters of the *Bhagavad Gita*, Krishna seeks to correct these and many other misunderstandings. Specifically regarding *varna*, Krishna states in *Bhagavad Gita* 4.13:

*I founded the four-varna system
with gunas appropriate to each.
Although I did this, know
that I am the eternal non-doer.*

Krishna thus refutes the idea that *varna* is based on birth, asserting instead that it is determined by an individual's *gunas*, or qualities. Approximately twenty percent of the *Bhagavad Gita* is devoted to explaining the three *gunas – sattva, rajas,* and *tamas* – and how they shape an individual's personality, and, consequently, *varna*. By devoting considerable attention to this subject, Krishna emphasizes the importance of dispelling misconceptions about the *varna* system, including the belief that it is determined by birth.

Classifying People

The most important statement of Hinduism is that *Brahman* is within us. Our purpose in life is to demonstrate our own divinity in thoughts and actions and to see this divinity in all. The Hindu greeting of *namaste*, with folded hands, is greeting the common divinity in others. Hinduism exhorts us to see this unity in diversity.

Is it then appropriate to classify people? The answer is yes if the purpose is right. The path to unity is through diversity.

Consider blood type as one example. While everybody has blood that is liquid and looks red, there are four different common blood groups – A, B, AB, and O. There is an Rh factor, the presence or absence of which makes the eight most common types of blood. There are specific ways in which blood type must be matched for safe blood transfusion; otherwise, a patient's immune system may attack the transfused blood. It is appropriate to classify people by their blood type.

There are multiple ways in which we classify people for various purposes – by external characteristics, by physical body, by mind and intellect, and by personality. We have control over some classifications, such as body weight, that we can improve. We do not have control over other classifications, such as blood type. Every classification has its advantages, but every classification can also be misused. If the good purposes outweigh the bad, the classification is deemed useful.

External characteristics: We classify people by external characteristics including the country they are from, the language they speak, the religion they belong to, their race, and their education.

Physical body: We classify people as men and women. We classify people by their height, weight, and BMI. We classify people by the size of their feet, the circumference of their neck, and the color of their skin, among other measurements.

Mind and intellect: We classify people by their emotional quotient (EQ), and their intelligence quotient (IQ).

Personality: More recently several tests have been developed to classify people by their personality. For example, one test classifies people in four personality types – A, B, C, and D. Individuals with Type A personality are goal-oriented, risk-takers, and good under stress. Those with Type B personality are relationship-oriented, outgoing, and enthusiastic. Individuals with Type C personality are meticulous, logical, and prepared, while those with Type D personality are task oriented, stabilizing, and cautious. Each personality type has its strengths and weaknesses, its suitability for certain work, and its motivations. None of the personality types is necessarily deemed better than the other.

Several thousand years ago, *Sankhya* philosophers effectively proposed classifying people based on their concept of causal body. The causal body controls our tendencies and thereby influences our body, mind, and intellect. Our thoughts and desires, the way we talk, act, and behave, the work we can happily perform are all influenced by our tendencies, governed by our causal body. Causal body influences our physical, emotional, and intellectual personality. Classifying people by the perceived composition of their causal body is a more fundamental approach to classification compared to classifying people by their body, mind, and intellect.

Varna System

The *Chatur Varna* system classifies people into four functional classes suitable to serve specific needs – *Brahmin, Kshatriya, Vaishya,* and *Shudra*. Though its earliest mention is cosmological rather than sociological, later philosophical traditions, including insights from the *Bhagavad Gita*, interpret these groupings as reflective of the interplay of the three *gunas – sattva, rajas,* and *tamas* – which constitute *Prakriti* or primordial nature. While there is no direct evidence that early *Sankhya* philosophers explicitly systematized this social structure, inherent tendencies offer a compelling framework for understanding *varna* as a dynamic, functional organization rather than a rigid hierarchy. The concepts of *Prakriti* and *gunas* have been discussed in detail in previous chapters, and we briefly revisit them here.

Prakriti

Prakriti is an independent ultimate reality – the subtlest, unconscious form of primordial matter – the source from which the universe evolves through continuous transformation. According to *Sankhya* thought, *Prakriti* is composed of three *gunas*: *sattva, rajas,* and *tamas*. These are not substances but fundamental qualities. Everything in the universe – animate and inanimate – arises as a function of these *gunas*.

To make the metaphysical concept of *Prakriti* accessible for practical purposes in this book, we describe *Prakriti* as a combination of three *gunas* that add to 100%. This simplified model is useful for profiling personalities, tracking spiritual progress, and mapping tendencies (*varna*).

An analogy from modern physics helps illustrate this idea. Just as all elements in the periodic table arise from varying numbers of three basic subatomic particles – protons, neutrons, and electrons – so too do all phenomena in the universe arise from varying proportions (not quantities) of the three *gunas*. While physics emphasizes particle count, our simplified model emphasizes the percentages of *gunas*.

Gunas

The *Bhagavad Gita* elaborates on how the *gunas* manifest in human behavior:

- *Sattva* is the *guna* of clarity, balance, and light. It manifests as awareness, tranquility, compassion, and the pursuit of truth and wisdom. A *sattvic* life is contemplative and harmonizing.
- *Rajas* is the *guna* of activity and passion. It leads to ambition, desire, restlessness, and domination. A *rajasic* life is outwardly engaged and often driven by personal gain.
- *Tamas* is the *guna* of inertia and obscurity. It gives rise to confusion, apathy, lethargy, and ignorance. A *tamasic* life is dull, inactive, and resistant to change.

These qualities fluctuate in all individuals across time and situations. When we speak of someone being predominantly *sattvic*, we mean their average disposition over time.

Varna

Our causal body, consisting of unique percentages of the three *gunas* that add to 100%, influences our *vasanas* (tendencies), and thereby our thoughts, desires, and actions. This results in our unique personality. The three *gunas* are like the three primary colors. Just as mixing the three primary colors red, yellow, and blue in different proportions creates an infinity of hues; changing the proportion of the three *gunas* results in an infinity of human personalities. In this light, the four *varnas* may be viewed as broad personality types:

- *Brahmins*: Predominantly *sattvic* – introspective, wise, and contemplative.
- *Kshatriyas*: Predominantly *rajasic* – courageous, active, and justice-driven.
- *Vaishyas*: Balanced mix – practical, enterprising, and socially adaptive.

- *Shudras*: Predominantly *tamasic* – supportive, grounded, and comfort-seeking.

This classification is not based on birth, occupation, appearance, or even intellect, but on *vasanic predisposition* and *tendencies* – a deeper layer of identity. It is meant as a guide for self-awareness, improvement, and fulfillment rather than a tool of exclusion.

A Note of Caution
Every classification can be misused. While tendencies may align with certain roles – just as height may influence basketball aptitude – this framework should never be used to limit opportunities, deny education or employment, or restrict social mobility. The *varna* assessment is intended as a tool for self-understanding, improvement, and personal fulfillment, not as a determinant of what someone can or cannot do. Everyone should remain free to pursue any path regardless of their assessed *varna*. Personality assessments can change over time or be incorrect. This philosophical framework for understanding human tendencies is fundamentally different from the caste system and must never be used to justify discrimination or social stratification. The goal is to help individuals find work and life paths that bring them greater satisfaction, while keeping all doors open to all people.

Mathematics of *Varna*

Everyone has a causal body consisting of a unique proportion of the three *gunas*. This proportion has a large influence on an individual's personality and suitability for performing certain societal functions. Two questions arise:

1. How many classes (*varna*) should individuals be grouped into?
2. What should be the *guna* composition of each class?

For practical reasons, the number of classes should be small. It would have been easy to propose three classes, calling them *sattvic* (S), *rajasic* (R), and *tamasic* (T), and suggesting that a *sattvic* individual has a large proportion of *sattva*, a *rajasic* individual has a large proportion of *rajas*, and a *tamasic* individual has a large proportion of *tamas*. Six classes could have been proposed by considering the three *gunas* in descending order: SRT, STR, RST, RTS, TSR, and TRS. There could have been seven classes by adding a class called *equal gunas*. But this is not what was proposed.

Mathematical Equation for Causal Body.
Causal body contains the unique proportion of three *gunas* that everyone has. We now answer the above two questions by proposing a mathematical equation to represent causal body. We have previously assumed that the three *gunas* always add to 100% – as a useful simplification to profile personalities, track spiritual progress, and map tendencies.

In the context of causal body, this assumption may be restated as follows:

At any point in time, everyone has a causal body consisting of a unique blend of the three gunas, which together sum to 100%. If gunas are expressed as a fraction, their sum is 1.

Mathematically this is represented by the following mixture equation.

Causal Body = α *Sattva* + β *Rajas* + γ *Tamas* where $\alpha + \beta + \gamma = 1$ or 100% ……. (8.1)

Each of the three coefficients in the equation represents the fraction or percentage of that *guna*. The constraint is that the three coefficients in Equation 8.1 must add up to 1 or 100%. If $\alpha = 1$ then β and γ must be zero, and the person is 100% *sattvic*. If α, β and γ are equal, each is $1/3^{rd}$, and the individual has a balanced combination of three *gunas*. Another individual could be 20% *sattvic*, 60% *rajasic* and 20% *tamasic*. The percentages must always add to 100%.

Triguna Tricone

Figure 8.1 illustrates the three *gunas* graphically plotted along three independent axes. Due to the constraint that the three *gunas* must add to 100%, their percentages cannot be arbitrarily selected. For instance, it is impossible to have 100% *sattva*, 100% *tamas*, and 0% *rajas* simultaneously because the three *gunas* must add to 100%, not 200%. The region where the three *gunas* add to 100%, known as the feasible region, forms the shaded triangle shown in Figure 8.1, and the causal body is constrained to be within this triangle. We refer to this triangle as the *Triguna Tricone* – the triangle of three *gunas*.

Figure 8.1: Feasible Region for the Causal Body (*Triguna Tricone*)

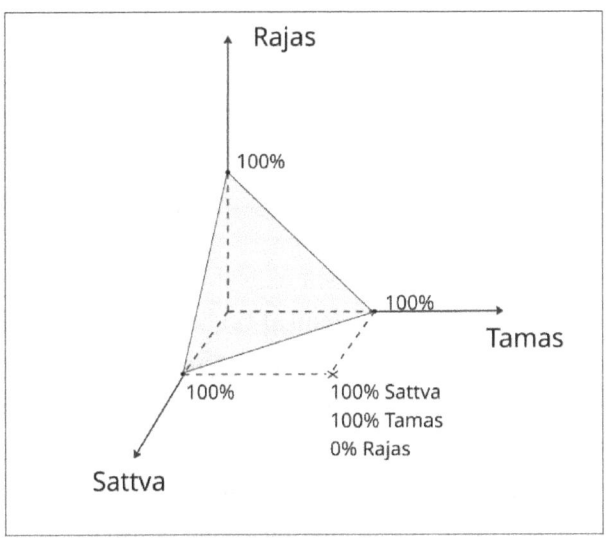

Figure 8.2 provides a detailed view of the *Triguna Tricone*. The vertices of the triangle represent 100% pure *gunas*. For example, the topmost vertex of the triangle is 100% *rajas*. The *rajas* axis extends perpendicularly down from the top vertex. Where it intersects the lower horizontal line, *rajas* is 0%. The various horizontal lines indicate how *rajas* increases from 0% to 100% in increments of 20%. The other two axes are similarly constructed. In Figure 8.2, Point 1 represents 80%

rajas, 10% *sattva*, and 10% *tamas*. Point 2 is in the middle where the three axes intersect and represents *gunas* in equal proportion.

Figure 8.2: *Triguna Tricone*

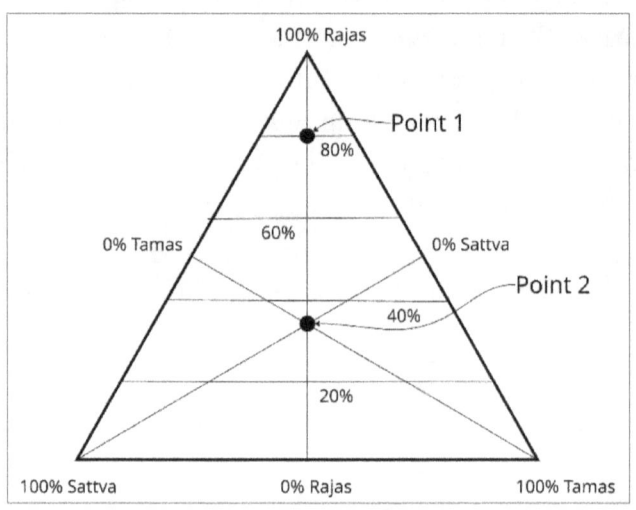

Chatur Varna – **Four Personalities**

Every individual on this earth has a causal body consisting of a unique mixture of the three *gunas*. The causal bodies of all individuals are inside the *Triguna-Tricone* (triangle of three *gunas*) shown in Figure 8.2. Even though everyone is unique, for practical purposes we need to group all individuals into a few categories. People's tendencies are a function of their causal body. The categories should be such that the tendencies of people within each category are similar, and those in different categories are substantially different. The categories together should cover the entire triangle. How many categories should there be?

Figure 8.3 shows what will happen if we use two, three, and four categories.

If the triangle is split into two parts, as in Figure 8.3 using (say) the vertical line in the middle, then the two categories will consist of the left and the right portions of the triangle. Each category will include people with widely different tendencies – ranging in the left portion of the triangle from 100% *sattvic* to 100% *rajasic* to an equal proportion of all *gunas*. This would defeat the purpose of categorization.

Using three categories, as in Figure 8.3, still results in a wide variation in the proportion of three *gunas* within each category.

The smallest practical number of categories is four, as shown in Figure 8.3, corresponding to predominantly (meaning greater than 50%) *sattvic* tendencies, predominantly *rajasic* tendencies, *balanced* tendencies (where no *guna* exceeds 50%), and predominantly *tamasic* tendencies. In Figure 8.3, these are respectively denoted by S, R, B, and T personalities, namely, *sattvic, rajasic, balanced,* and *tamasic personalities*. These respectively correspond to the traditional *Brahmin, Kshatriya, Vaishya,* and *Shudra varnas*.

While the triangle could be subdivided into larger number of categories, for example, nine categories or nine equal triangles by dividing each of the three axes into three parts, nine categories are too many for practical purposes. The uncertainty in judging the percentage of three *gunas* in an individual is too large to justify these many classifications.

The reasons for the four *varnas* and their *guna* composition follow directly from Equation 8.1, which is a mathematical statement of a key assumption of *Sankhya* philosophy. Although the original systematization may not have been done this way, the mathematical approach explains and provides justification for the four *varnas* and their composition.

Figure 8.3: Four types of personalities

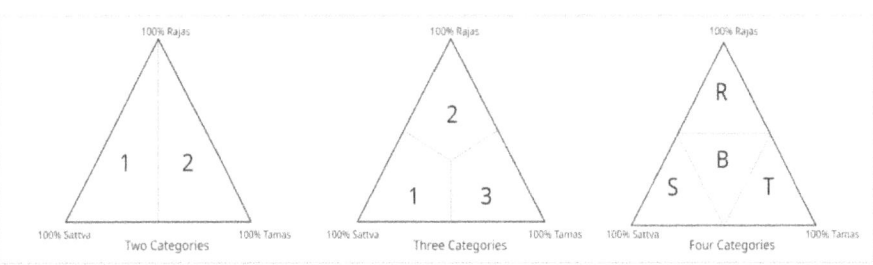

What Color is Your Personality?

Unlike height and weight, causal body cannot be measured directly. Emotional quotient (EQ) and intelligence quotient (IQ) also cannot be measured directly and are assessed indirectly through structured tests. Such is also the case for the various personality assessment tests available today. *It is necessary to develop a similar test to measure our PQ – personality quotient – based on the three gunas.* Some of the evaluation criteria to construct such tests are described below.

Evaluation Criteria

In Chapter 18, verses 7 – 39, the *Bhagavad Gita* provides guidance to evaluate one's personality. These verses classify knowledge, action, actor, intellect, steadfastness, renunciation, and happiness under the three *gunas*. This framework enables individuals to assess their personality, and identify opportunities for self-development.

Knowledge: *Sattvic* knowledge sees unity in diversity and fosters equanimity and inner peace. *Rajasic* knowledge focuses on superficial diversity, leading to separation and comparison. *Tamasic* knowledge is narrow, irrational, and clings to partial truths.

Action: *Sattvic* action is dutiful, free from attachment, and performed with serenity. *Rajasic* action is ego-driven and result-oriented, generating anxiety. *Tamasic* action is impulsive, careless, and undertaken without regard to consequences.

Actor: *Sattvic* actor is detached, steadfast, and composed in all outcomes. *Rajasic* actor is driven by desire, prone to elation and despair, and often disregards ethics. *Tamasic* actor is lethargic, irresponsible, and indifferent to duty.

Intellect: *Sattvic* intellect discerns truth from falsehood and upholds *dharma*. *Rajasic* intellect misjudges right and wrong, and is guided by personal bias. *Tamasic* intellect is deluded and unable to make sound judgments.

Steadfastness: *Sattvic* firmness supports one's progress on a spiritual path. *Rajasic* firmness clings to self-imposed obligations, often rooted in fear or ambition. *Tamasic* firmness perpetuates fear, grief, or despair through stubborn inaction.

Renunciation: *Sattvic* renunciation is a release of attachment to results, not of responsibility. *Rajasic* renunciation avoids difficult duties out of discomfort. *Tamasic* renunciation stems from ignorance and laziness.

Happiness: *Sattvic* happiness arises from inner clarity and self-discipline. It may feel effortful at first but becomes enduring. *Rajasic* happiness is born of desire and sensory pleasure – initially enjoyable, ultimately unsatisfying. *Tamasic* happiness is rooted in ignorance and leads to harmful consequences.

Self-Assessment and Growth

These criteria allow individuals to reflect on their own tendencies across key dimensions. By examining their thoughts, desires, and actions, they can begin to identify their prevailing guna orientation and determine which *varna* they presently align with –*Brahmin, Kshatriya, Vaishya,* or *Shudra*.

This assessment is not a judgment but a stepping-stone toward personal transformation. As discussed in the essay "Theory of Reincarnation," the causal body evolves across lifetimes. One may begin this life with certain tendencies (Point 2 in Figure 8.2) and, through introspection and effort, shift to a different state over lifetimes (Point 1). This evolving causal body carries forward into subsequent births, gradually refining itself until liberation is attained.

The *varna* system, understood in this light, is a flexible and universal model for personal development. Like modern frameworks – such as the Type A/B/C/D classifications – it offers insights that cross cultural boundaries, while remaining deeply rooted in inner awareness.

To make this ancient framework practical, future work should focus on developing a standardized self-assessment tool. This includes constructing meaningful questions, offering real-world examples, guiding interpretation, and presenting case studies of individuals who have used the *gunas-based* model for growth. Such efforts could transform philosophical insight into a pathway for self-understanding and fulfillment.

Transcending *Gunas*

One purpose of life, it is said, is to transcend *gunas* and realize unity with *Atman*. Practically, this means achieving inner freedom – liberation from suffering and the tendencies that bind us to mental restlessness and worldly entanglement.

What does it mean to transcend *gunas*?

It is appropriate to progress from *tamasic* to *rajasic* to *sattvic* tendencies. However, transcending *gunas* does not mean that one goes from *tamasic* to *rajasic* to *sattvic guna* and then crosses some imaginary line to go beyond *gunas*. In the game of American football, it is not like a running back who starts at the goal line, crosses the 30-yard line going from the *tamasic* to the *rajasic* zone, then accelerates and crosses the 80-yard line going from the *rajasic* to the *sattvic* zone, and ultimately scores a touchdown, thus going beyond the *gunas*. It has been shown above that *gunas* are not a sequence of lines but are constrained to be inside a triangle. Transcending *gunas* also does not mean reaching a *sattvic* state and then going outside the triangle.

Each *guna* binds differently. *Sattva* to worldly knowledge and happiness, *rajas* to action and desire, and *tamas* to ignorance and inertia. When governed by these tendencies, our thoughts and actions arise reactively rather than freely. Our actions may not reflect what is right, but merely what our conditioning dictates.

As shown in Equation 8.2 below, the actions we take are a function of the situation we are in (destiny), our mixture of *gunas* (causal body/tendencies), and our free-will (intellect). Transcending *gunas* involves shifting this balance so that action is no longer dictated by tendencies but by clarity and inner freedom.

Actions = f(destiny, causal body/tendencies, intellect) ……(8.2)

Bhagavad Gita (14.22) guides this transformation, teaching us to cultivate a state of equanimity and detachment from the qualities of *sattva, rajas*, and *tamas*.

> *Whichever qualities arise*
> *Sattva, rajas or tamas*
> *He hates not their presence*
> *Nor long for them when absent*

This teaching reveals two paths. The first – eliminating all *gunas* to reach a state of zero tendencies – is theoretically appealing but impossible, as the *gunas* are always present, and will always add to 100%.

Therefore, our focus shifts to the second approach. Accept the presence of these *gunas* while ensuring that our actions are guided by intellect and free will, rather than by inherent tendencies. This requires complete control over our tendencies, using our free will and intellect. Control of tendencies varies in difficulty; for example, overcoming addiction may be more challenging than overcoming laziness. Achieving this control becomes easier in a *sattvic* state, where clarity and purity of mind facilitate self-effort and discipline.

One who attains this state – acting without being swayed by *guna*-based impulses – is known as a *jivanmukta*, a liberated being. This state of liberation is marked by the ability to act rightly without the sway of tendencies, reflecting true mastery over oneself. Thus, transcending *gunas* is not about eliminating them but about rising above their influence through self-effort, intellect, and a *sattvic* way of living.

Practical Applications

The *three-guna* framework presented in this essay offers a sophisticated tool for personal development and organizational effectiveness that bridges ancient wisdom with contemporary needs. For individuals, understanding one's *guna* composition provides valuable guidance for career decisions and personal fulfillment. Someone with predominantly *sattvic* tendencies might thrive in roles requiring deep thinking, counseling, research, or spiritual guidance, while those with *rajasic* dominance may excel in competitive business environments, leadership positions, emergency services, or entrepreneurial ventures. Individuals with balanced *guna* compositions could prove invaluable in roles requiring adaptability and the ability to work effectively across diverse teams and situations. Those with predominantly *tamasic* tendencies could find satisfaction in roles requiring patience, stability, and careful attention to detail – such as skilled craftsmanship, maintenance work, quality control, manufacturing, or service positions that provide essential foundations for society's functioning. Rather than limiting choices, this self-awareness enhances decision-making by helping people understand their natural inclinations while keeping all paths open for exploration.

In organizational contexts, this framework offers powerful insights for team composition, management strategies, and conflict resolution. Understanding that employees have different motivational drivers and working styles can dramatically improve workplace dynamics. A manager might assign detail-oriented, methodical projects to those with stronger *tamasic* tendencies, while giving innovation-focused, high-pressure assignments to more *rajasic* team members. This approach recognizes that diversity of temperament strengthens organizations rather than creating hierarchy based on perceived superiority of one type over another. The framework also provides a structured approach to personal growth, helping individuals identify specific areas for development while appreciating their unique contributions to society.

Most significantly, this system offers a pathway to transcending reactive behavior patterns through increased self-awareness. By understanding how *guna-based* tendencies influence thoughts and actions, individuals can develop greater conscious choice in their responses to life's challenges. The practical wisdom lies not in eliminating natural tendencies but in achieving enough self-mastery to act from clarity rather than compulsion. Like understanding one's learning style or emotional intelligence patterns, *guna* awareness becomes another valuable tool for navigating relationships, career transitions, and personal development with greater intentionality and success. This ancient framework, when properly understood as descriptive rather than prescriptive, provides timeless insights into human nature that remain remarkably relevant for contemporary personal and professional development.

Summary

This essay presents a mathematical framework for understanding the ancient Hindu *varna* system through the lens of *Sankhya* philosophy's three *gunas* (*sattva, rajas,* and *tamas*). *Bhagavad Gita* suggests that the original *varna* classification was based on personality tendencies rather than birth, distinguishing it from the discriminatory caste system that emerged later. Using a mixture equation where the three *gunas* must sum to 100%, the essay demonstrates why four personality categories represent the optimal practical division: *sattvic* (*Brahmin*), *rajasic* (*Kshatriya*), balanced (*Vaishya*), and *tamasic* (*Shudra*).

The mathematical approach proves that all personality combinations lie within a triangular feasible region called the "*Triguna Tricone*." It is shown that while other numbers of categories are mathematically possible, four categories best balance the need for meaningful distinctions with practical simplicity, given the inherent uncertainty in assessing personality traits. Evaluation criteria drawn from the *Bhagavad Gita* provide a framework for self-assessment across dimensions like knowledge, action, intellect, and happiness, each manifesting differently under the influence of different *gunas*.

The essay emphasizes that this framework should serve as a tool for self-understanding and personal development rather than social stratification. It includes explicit warnings against using the system to limit opportunities or justify discrimination, positioning it instead as comparable to modern personality assessments. The goal is to transcend categorization entirely – achieving liberation through conscious choice rather than reactive behavior driven by inherent tendencies. The practical applications include career guidance, team building, and personal growth, offering a bridge between ancient wisdom and contemporary organizational and individual development needs.

CHAPTER 9
STHITAPRAJNA AND ROBUST DESIGN

Can the engineering principles behind robust product design illuminate the path to inner peace? What if the secrets of tranquility lie not only in ancient philosophy, but also in modern engineering?

In this thought-provoking essay – part dialogue, part philosophical inquiry – an engineer and a statistician unpack a surprising parallel between Sthitaprajna, the ideal person of the Bhagavad Gita, and robust product, the ideal of engineering design. By comparing how systems withstand disturbances with how minds cultivate equanimity, the essay reveals a shared mathematical structure behind stability – whether in matter or in mind. Readers will discover that engineering resilience and spiritual wisdom share a common blueprint.

A *Sthitaprajna*, as described in the *Bhagavad Gita*, embodies the ideal of steady wisdom – a person fully engaged with the world, with a mind unperturbed by life's ups and downs. Similarly, a robust product represents the pinnacle of engineering design – it functions at optimal level regardless of disturbances. One belongs to the realm of spiritual self-mastery, the other to mechanical resilience. Yet, there is a striking conceptual parallel between the two. Neither has withdrawn from the world; both remain actively involved and operate at their peak, unwavering in the face of causes beyond their control.

This essay explores how the mathematical principles that ensure the robustness of inanimate systems can illuminate the early stages in the transformation of a human being into a *Sthitaprajna*. Through this comparison, we uncover a shared logic of equilibrium: harmony achieved not by retreat but by resilience.

Drug Transfer through Skin

The fact that people's skins are different is a problem, the Engineer thought. He was not thinking about the color of people's skins. He had just attended a lecture where the speaker described the many ways in which people differed with respect to their skins. Some were thick skinned; others were thin skinned. Some wore old and wrinkled skins; others had new and tight skins. Some had hairy skin, and others had skin smooth as silk. Some people had oily skin, others did not. The number of pores per unit area also varied. Even for the same person the skin on different parts of the body was different. This variation influenced drug transfer through skin.

The Engineer had been named a member of the cross-functional team set up to develop a new drug-containing patch to transfer drug through skin. The idea was to achieve a desired constant rate of drug transfer, called flux, for each patient. There were many potential benefits, not the least of which was the fact that continuous constant dosage meant that less overall dose was required while achieving improved efficacy and reduced side effects.

The cross-functional team was set up to look at the problem of developing such a product from all perspectives that could come into play throughout the life cycle of the product. Even though the product was in early development, the company wanted many of the downstream issues to be resolved as early as possible in the development process. The company had been burnt badly before, when product design issues had been unknowingly handed off to manufacturing, and at the manufacturing stage, the solutions had been extremely expensive.

The Engineer had observed how the product-development scientists and engineers did their experiments. The scientists would screen different compounds and their amounts to develop just the right formulation. The engineers were experimenting with bench scale processes and changing the processing conditions to determine their effects. For each formulation or process change they needed to know what the rate of drug transfer was and how it varied over time. This could only be measured using cadaver skins. Because the cadaver skins varied so much, scientists had gone to great lengths to identify skin samples that were as alike as possible, so that they could understand the effects of the formulation and process changes without the results being vitiated by the skin-to-skin differences.

The Engineer thought that this way of developing products was problematic. If the effect of skin-to-skin differences was not purposefully considered in these experiments, and if ways of counteracting the effect of skin-to-skin differences were not found at this early stage, then a product and process could be settled upon which was overly sensitive to skin-to-skin differences. This would lead to a different rate of drug transfer for different patients, degrading efficacy. A different product would have to be designed for people with significantly different skins and the multiplicity of products would cause chaos.

The Engineer wanted to discuss the subject with his friend the Statistician. They met at a local restaurant for dinner. It had become their favorite place to meet. It gave them a chance to unwind, and discuss both work-related and non-work-related topics.

Karma Yoga

The Statistician was in a philosophical mood. He had been attending lectures on *Bhagavad Gita*, the "song of God," the famous book of Hindu philosophy that explained the way to *Karma Yoga* and becoming a *Sthitaprajna*.

"Two of the most quoted stanzas in Gita explain *Karma Yoga*" the Statistician began. Here is what they say:

> *You have control over your actions,*
> *but not on their fruits.*
> *Act for action's sake,*
> *do not be attached to inaction.*
>
> *Perform actions in this world,*
> *abandoning attachments.*
> *Be alike in success or failure,*
> *Yoga is perfect evenness of mind.*

"The stanzas teach us to perform actions in this world without attachment to results, because the outcomes are not entirely in our control. It goes without saying that we should perform thoughtful and skillful actions – if we are not a surgeon, we should not perform surgery. As we go through life, many factors – social, economic, emotional, physical, even traffic jams while going to work – can affect our mental balance and happiness. We should condition ourselves, you may even say, we should redesign ourselves, to keep evenness of mind in the face of all these disturbances. Such a person is called a *Sthitaprajna.*"

"Are you a *Sthitaprajna?*" the Engineer asked with a twinkle in his eye. "No" replied the Statistician with a smile. "I always say – Do as I say, not as I do."

The next two sections involve some mathematics. If that is a concern, you may go directly to the section titled "Beyond Robustness: The Complete Sthitaprajna."

Mathematics of Robust Design

"Let me tell you about this project I am working on" the Engineer said, changing the subject. He briefly described the project. "What we really need is a way of designing our drug-transfer patch so that any skin-to-skin differences will have no effect on drug transfer" he said. "I could even call such a product, as you say, a *Sthitaprajna* product."

The Statistician was impressed with the ability of the Engineer to connect disparate ideas. "There is a method to do that" the Statistician said. "It is called Robust Design. The idea of robustness and the idea of *Sthitaprajna* are similar. Factors like skin are called noise factors because they are not in the hands of the product designer to control. Much like what your coworkers say about you is not entirely in your control and yet you want to maintain evenness of mind regardless of their opinion. Here, we want to maintain evenness of drug transfer even when the skin varies."

"Can you explain the method to me?" asked the Engineer.

"There is a lot to this method, and I cannot explain all the intricacies in a short time," said the Statistician. "Let me explain one key idea."

The Statistician took out a piece of paper and a pencil from his pocket. He was in the habit of carrying these items. He thought that he was beginning to forget things, and there was no telling when a bright idea might strike.

"Look at it this way" he continued. "Let us say that A is a formulation factor, the amount of some ingredient in your product. We call it a control factor because you can control how much of that ingredient to add, much like you can control the way you think if you put your mind to it. Suppose we do experiments at two levels of A, a low level of 10 mg. and a high level of 20 mg. We will designate these levels as -1 and $+1$. N is a noise factor, in your case, skin. We call it a noise factor because it is beyond our control in actual usage. But we can control it just for the purpose of experimentation. Suppose we have two levels of N. The -1 level could be a porous, thin cadaver skin, and the $+1$ level could be a dense, thick cadaver skin. This captures the range of skins your product may see in practice. Suppose we conduct the four experiments corresponding to the four combinations of the levels of A and N and measure the flux. The collected data and the plot may look like this" he said, pointing to Figure 9.1.

Figure 9.1: First set of results

A	N	Flux
-1	-1	10
-1	+1	30
+1	-1	15
+1	+1	15

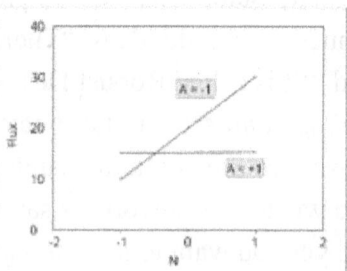

"What this means is that when ingradient A is − 1 (10 mg.), skin (noise factor N) has a big effect on flux and flux changes from 10 to 30 depending on the skin type. When ingredient A is +1 (20 mg.), skin has no effect on flux; the flux stays constant at 15. The graph illustrates how the effect of noise factor N changes as the level of control factor is changed. When the effect of a noise factor changes as a function of the level of control factor, we call it interaction between control and noise factors. *This interaction between a control factor and a noise factor is the key to robustness.* If we cannot find a control factor that interacts with skin, we are out of luck on making your patch robust. Do you understand?"

"I understand," said the Engineer. "I can see that if A = + 1, the effect of skin disappears. For all different skins, I have the same flux, meaning that the patch is robust. All individuals will have the same rate of drug transfer regardless of their skin. If the two lines for A = -1 and A = +1 were parallel with a slope, there would be no interaction between the control factor A and the noise factor N, meaning effect of N would be the same at both levels of A. In this case, I could not achieve robustness using control factor A. Is that right?"

"Exactly," said the Statistician. "You need an interaction between a factor you can control and a factor you cannot control to achieve robustness."

"But if A = + 1, then the flux will be 15. What if the desired target flux is different from fifteen?" asked the Engineer.

"Can you think of factors other than A that will change the flux?" asked the Statistician.

"There are many such factors" the Engineer replied. "I can see that we could use another factor that changes the flux but does not interact with the noise factor (skin) to get the desired flux, while using control factor A to achieve robustness."

"There is more to robust design than I have said so far. When your team meets to discuss the problem, call me in. I will explain the experiment design and analysis procedure" the Statistician said.

"That sounds good," said the Engineer. But he remained deep in thought for several minutes before speaking again.

"I like the idea," he said. "What the scientists need to do is to explicitly introduce the noise factor in their experiments. They should experiment in a way that allows control-by-noise interactions to be determined, and see if they can counteract noise by the proper selection of control factors and their levels. This is not happening now because they are not experimenting with widely different skins. It needs to happen. But the flux we get under the four experimental conditions in your table is not in our control. Isn't that what *Bhagavad Gita* says? Do what is right, but the results are not entirely in your hands? What if the results turned out this way?" He changed a few numbers in the table and drew a new graph shown in Figure 9.2.

Figure 9.2: Second set of results

A	N	Flux
-1	-1	10
-1	+1	30
+1	-1	30
+1	+1	10

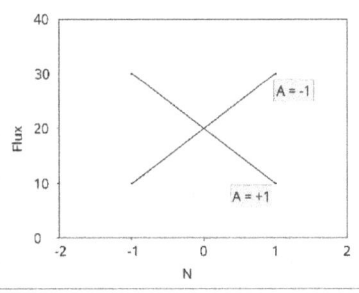

"You can see from the graph that there is still an interaction between A and N. The effect of noise on flux is completely different for the two levels of A. We should be able to use A to achieve robustness. How do you find the right value of control factor A in this case?"

"We can model the interaction between control factor A and noise factor N explicitly. We can build an equation for flux like this." The statistician did some calculations and wrote the following equation.

$$\text{Flux} = 20 - 10 \, AN \quad \text{...(9.1)}$$

"When $A = -1$ and $N = -1$, then $AN = +1$ and Flux = 10 which matches the observed flux in the table. When $A = -1$ and $N = +1$, $AN = -1$ and from the equation, Flux = 30, which also matches the observed Flux. And so on. The equation fits the data. It is obvious from the equation that when control factor $A = 0$, the effect of N disappears. So, in this case, the amount of ingredient A must be at the middle level, designated by zero. In this example, $A = 0$ corresponds to the middle level of the ingredient, which is 15 mg. You can see this from your graph as well. As A goes from -1 to $+1$, the line will rotate clockwise and when $A = 0$, it will become horizontal, and at that point, the effect of skin will disappear."

Mathematics of *Sthitaprajna*

Everyone seeks happiness, often finding it in the fulfillment of desires. However, this happiness is fleeting. Achieving our desires is not always within our control, and despite our best efforts, we may face disappointment, leading to unhappiness. The pursuit of desires does not guarantee even momentary happiness, let alone a lasting one.

Another perspective on happiness is viewing it as a state of mind. When our mind is agitated, we experience anxiety, fear, and sorrow. Conversely, a calm mind brings joy. Therefore, happiness can be measured by the equanimity of our mind.

In our daily lives, we encounter various situations—physical, emotional, economic, and social. Victory brings joy, while defeat brings dejection. Our mind oscillates between happiness and unhappiness. When we lack control over our senses and mind, our pain can turn into suffering, diminishing our quality of life.

A *Sthitaprajna* is a wise individual with an even mind, unperturbed by life's ups and downs. This does not mean withdrawal from the world or avoiding life's challenges. A *Sthitaprajna* engages with the world fully, without being callous, indifferent, or devoid of feelings. Instead, he has total control over his senses and does not let his thoughts and actions be unduly influenced by the circumstances he encounters. He feels emotions, without being emotional.

Problem Formulation

The benefits and ways of achieving equanimity in the presence of external and internal disturbances, originally stated millennia ago, may be restated in a format like the one used to design a robust product. We begin by defining responses, noise factors, control factors, and ways to reduce the effects of noise.

Responses: The output(s) we seek is called response – in this case the magnitude of mental perturbations. We seek to reduce these perturbations and achieve evenness of mind or peace of mind.

Noise factors: These are obstacles to achieving equanimity, called noise factors, which are not in our control. They are categorized into two types – external and internal. In each case, the noise factors may be beyond our control, or expensive to control. This leads to four possibilities:

1. External noise factors, beyond our control: Examples include what other people say about us, unexpected physical and mental violence, accidents, and natural disasters.
2. External noise factors, expensive to control: Examples include ambient conditions (e.g., temperature, and humidity), and economic losses.
3. Internal noise factors, beyond our control: Examples include birth defects and age-related factors.

4. Internal noise factors, expensive to control: Examples include physical and mental sickness, diseases, and addiction.

Control factors: These are factors within our control, actions we can take, using our body, mind, intellect, and tendencies. As examples, they include sense control, mental attitude, and physical conditioning.

Ways to reduce effects of noise factors: Noise factors cause perturbations in our mind and prevent us from achieving evenness of mind – from being a *Sthitaprajna*. There are two ways to reduce the effect of noise factors:

1. Eliminate or control the noise factors themselves.
2. Let the noise factors exist, but reduce or eliminate their effects by finding control factors that interact with them.

It is the role of science, technology, and socioeconomic, legal, and political systems to eliminate or control the noise factors. For example, ambient temperature is a noise factor and can be a source of considerable dissatisfaction as it varies from below zero to above 100 degrees F. Heating and air conditioning technologies now allow people to live comfortably at 72 degrees F regardless of the outside conditions. Medical science has made great progress in eradicating and curing diseases. Banning harmful substances, increasing the price of cigarettes, and providing social safety nets are examples of eliminating or controlling noise factors.

Despite progress, total elimination of noise factors may never be feasible. Some will persist causing pain and suffering. Therefore, to counteract the effects of noise factors, we must find control factors within our body, mind, intellect, and tendencies that interact with noise – that have the capacity to reduce the *effect* of noise factors.

The practical decision-making process is simple. For any noise factor affecting our equanimity, we must decide: can we do something meaningful about this factor or not? Whether the solution requires immediate action or future planning, involves personal effort or professional help, costs little or demands significant resources, the fundamental choice remains binary. If we can meaningfully and acceptably address the noise factor through external means, we may want to pursue that path. If we cannot – either because it is truly beyond our control or because we have decided the cost of control is prohibitive – then we must build internal robustness against it. This clear decision framework prevents us from getting stuck in endless analysis and moves us toward action.

Sthitaprajna Equation

Sthitaprajna is one who is not unduly excited by a pleasant environment or plunged into dejection by an unpleasant environment. One approach to try to accomplish this is to be like the tortoise who withdraws his limbs at the slightest sign of danger. *Bhagavad Gita*, verse 2.58 says:

> *Drawing back his limbs*
> *As a tortoise does from danger*
> *A man of wisdom*
> *Withdraws his senses from sense objects*

A person with control over his senses and actions could do the same – automatically withdraw his senses at the sight of temptation. This approach is wise but is an example of reducing or eliminating noise itself, not necessarily of building robustness against noise while experiencing the noise. An example is a person who is addicted to alcohol and is tempted to enter a bar he sees, but walks away. With practice, one may develop an attitude that removes the attraction of worldly pleasures, becoming free from likes and dislikes, thereby building robustness against noise. However, this approach risks withdrawing from the world.

Entering the bar and not drinking exemplifies building robustness against noise. *Bhagavad Gita*, verse 2.64 says:

> *But a self-controlled man*
> *Moving among objects*
> *Without craving or aversion*
> *Attains peace*

To build robustness against noise, we seek a control factor that interacts with noise, namely, a control factor that changes the effect of noise and has the capability to reduce its effect to zero. Tibetan monks present one example of using our internal control factors to counteract noise. They condition their body to be able to withstand a large variation in temperature, including very cold temperatures. Ambient temperature is a noise factor. The monks do not control this noise factor by living in an airconditioned environment, but rather use a conditioned body to dramatically reduce its effect. For them, the noise factor does not produce noise anymore! That is redesigning the human body to make it robust.

One could use many different control factors to build robustness against various noise factors. Our internal control factors are associated with our body, mind, intellect, and tendencies. Conditioning our body with appropriate exercise and keeping it fit is one example of using a control factor to reduce the effect of noise, such as the risk of falling and getting hurt. Controlling our thoughts and our responses to events is an example of using our intellect to reduce the effect of external noise. Ensuring that our tendencies, likes, and dislikes, do not drive our actions – transcending *gunas* – is another example of building robustness against noise.

A question arises: What is in our control that interacts the most with noise? What is in our control that can significantly reduce the effects of both internal and external noise factors? Our *attitude to life* is such a factor. Our attitude is simultaneously the easiest to change and yet so hard. While attitude change theoretically requires only a shift in mental perspective, deeply embedded conditioning and life circumstances often make such transformation a gradual process requiring patience and sustained practice. Depending upon our attitude we can view the same situation positively or negatively, happily or unhappily. Let

H = Happiness going from very unhappy (denoted by 0) to very happy (denoted by 1)
N = Noise going from no noise (0) to extreme noise (1)
A = Attitude going from negative (-1) to positive (+1)
H_{max} = Maximum happiness when there is zero noise (1)

As noise increases, happiness reduces, and this could be approximated by a straight line with a negative slope as follows:

$$H = H_{max} - \alpha \text{ Noise} \qquad \alpha \geq 0 \qquad \ldots\ldots\ldots\ldots\ldots (9.2)$$

Where α is the slope indicating the rate at which happiness reduces as noise increases. We want the slope α to be small, ideally zero. Then the effect of noise will disappear, and our happiness will stay at the highest level. For example, if noise happens to be a major illness, it will typically reduce our happiness. If α is large, a person may get depressed. If α is small, a person may be able to weather the storm.

The absolute value of α depends upon a person's attitude indicating an interaction between attitude and noise, namely, depending upon the attitude the effect of noise changes. A person with a negative attitude may go into depression at the slightest noise, meaning that when attitude A = -1, α is large. For a person with a positive attitude, when A = +1, α is small. This relationship between α and attitude may be approximated by:

$$\alpha = \beta_0 - \beta_1 A \qquad \beta_0 \geq \beta_1 \qquad \ldots\ldots\ldots\ldots (9.3)$$

As attitude (A) becomes more positive, α reduces. For A = +1, α = ($\beta_0 - \beta_1$) becomes the smallest possible, ideally zero. When α = 0, from Equation (9.2), happiness stays at the maximum level regardless of noise.

Elements of cultivating a positive attitude include self-awareness, non-attachment, discipline, and wisdom. Self-awareness is clearly understanding our own thoughts, emotions, and actions. Non-attachment is letting go of excessive attachments to material possessions and fruits of actions, and finding contentment within. Discipline is the ability to control our senses, impulses, and reactions. Wisdom is making decisions based on knowledge rather than emotions. A person with such a positive attitude, such tendencies, is a *Sthitaprajna*, an individual with evenness of mind – unperturbed by life's ups and downs.

The mathematics of designing a high quality robust inanimate product like the drug-transfer patch, and the mathematics of re-designing a living human being to be a *Sthitaprajna* are similar. Both require an interaction between something one can control, and something one cannot control.

Beyond Robustness – The Complete *Sthitaprajna*

The previous sections explored the impermanence of happiness derived from fulfilling desires, contrasting it with a deeper, more stable happiness rooted in mental equanimity, introducing the concept of Sthitaprajna – a wise person who remains unshaken by external events and internal emotions. Rather than withdrawing from life, such an individual engages fully with the world while maintaining complete mastery over the senses and reactions. The core idea is that true happiness arises not from favorable circumstances but from a tranquil, balanced mind.

The tranquility of our mind is disturbed by factors that are either not in our control (what people say about you, sudden death of a loved one, natural disasters such as a tornado), or are difficult or expensive to control (ambient conditions, traffic jam). While external systems like science, technology, and government policies can help reduce or eliminate of some of these factors, lasting peace demands internal robustness. This means developing control over our physical body, senses, thoughts, attitudes, habits, and responses that diminish the impact of factors outside our control – much like Tibetan monks who train their bodies to withstand extreme cold. This is building robustness and is an important step to becoming *Sthitaprajna*.

In the *Bhagavad Gita*'s second chapter, Krishna responds to Arjuna's question about the qualities of *Sthitaprajna* – the one of steady wisdom. This teaching, summarized below, represents one of the most practical and profound concepts in the entire text, offering a blueprint for achieving inner peace and stability in daily life.

1. *A true Sthitaprajna demonstrates complete sense control,* meaning he is not constantly pulled around by sensory desires or impulses. *He knows that from attachment to objects desire arises, from desire arises anger, from anger delusion, from delusion confusion of memory, from confusion of memory loss of intellect, and from loss of intellect he perishes (Bhagavad Gita 2.62-63).* This does not mean he avoids pleasures, but rather that he can enjoy good food without overeating, appreciate beauty without possessiveness, and experience life's offerings without being enslaved by the need for constant stimulation. He experiences emotions, without being emotional. He finds genuine satisfaction in moderation and develops the capacity to say no to immediate gratification when it conflicts with deeper values or long-term wellbeing. His intellect controls his mind.

2. *He remains unaffected by success and failure, exhibiting equanimity in all circumstances.* By following the principles of *Karma Yoga* (see Chapter 10), he performs duties without letting personal happiness depend on outcomes. His fulfillment comes from action itself rather than its rewards. This detachment does not make him cold or uncaring, but rather allows him to act more effectively because he is not clouded by selfish motivations or the anxiety that comes from being overly invested in results. He can love, work, and contribute to society while maintaining inner freedom from the need to control or possess.

3. *He is free from desires,* evolving from selfish pursuits to selfless actions guided purely by duty, eventually reaching a state of desireless action, where right action becomes second nature rather than a calculated choice. This freedom from desire and

attachment naturally leads to inner contentment. It is not passive resignation but an active state of fulfillment that remains stable regardless of what happens in the outside world.

4. *Self-knowledge forms the bedrock of his equanimity.* Just as a stable system resists perturbations, a Sthitaprajna resists the oscillations of suffering through alignment with the unchanging Self. Through sustained self-inquiry and spiritual practice, he understands the distinction between temporary experiences – emotions, circumstances, even physical sensations – and essential nature, which remains unchanged by external events. This is not merely an intellectual understanding but an experienced realization that provides unshakeable peace. When people truly know that their core being is not threatened by criticism, failure, loss, or even death, they naturally respond to challenges with wisdom rather than reactivity. This understanding gives them fearlessness and courage to face life's challenges, make difficult decisions, and stand up for what is right without being paralyzed by worry about consequences. They can act with conviction because they are not constantly protecting a fragile ego or trying to maintain a particular image.

In a world filled with uncertainty, emotional turbulence, and material distractions, cultivating steady wisdom has profound practical advantages. Wisdom in this case is not about intellectual mastery or expertise in every field. It is about remaining unshaken, discerning reality with clarity, and responding to life's trials without being overwhelmed. *Sthitaprajna's* wisdom transforms life's storms into mere ripples, ensuring that no external force can disturb the depths of inner peace.

Practical Starting Point

The ideal of *Sthitaprajna* – a person with unshakable wisdom and inner calm – is a high goal drawn from Hindu philosophy. But the method explained here offers a simple and useful starting point for anyone seeking more balance in everyday life.

Begin by checking in with yourself regularly. What has been disturbing your peace this week? Then ask honestly – can any of those problems be improved by outside help? For example, medical care for depression, financial insecurity, better sleep through changing your environment, support from friends or family, or using technology to manage stress. If you are dealing with serious mental health struggles, abuse, or injustice, or other situations requiring urgent external intervention, building inner robustness alone may not be enough – may even be counterproductive – and should not replace getting help or changing harmful situations.

For the things that truly lie outside your control, you can evaluate different internal strategies to see what helps reduce the noise and bring back calm. Think of this like running experiments. Emotional strength takes time to build, and what helps someone else might not work for you. Some people find that exercise helps them bounce back from stress. Others prefer meditation, creativity, or techniques like reframing their thinking – a glass half full rather than half empty. What matters is staying curious and methodical, not random – track what helps, instead of relying on what you *expect* should work.

What makes the above framework so powerful is how it brings *Bhagavad Gita*'s wisdom into real life. It shows that true strength is not about controlling the world, but about building inner skills to meet the world wisely. Whether you are managing tough relationships, health issues, money worries, or just the usual stress of daily life, this approach gives you a reliable path toward peace and effectiveness. Start by building this inner resilience, and over time, deeper qualities like wisdom, compassion, and contentment will naturally follow.

Serenity Prayer

There are similarities and differences between the practical advice offered above, and the advice offered by the Serenity Prayer. Here is the well-known Serenity Prayer:

God, grant me the serenity to accept the things I cannot change, the courage to change the things I can, and the wisdom to know the difference.

While both the Serenity Prayer and this essay begin with the same foundational wisdom – distinguishing between what we can and cannot control – they diverge dramatically in their prescribed responses to uncontrollable circumstances. The Serenity Prayer's approach involves identifying what cannot be changed, and finding peace by accepting these unchangeable realities. The *Sthitaprajna* framework takes a radically different path: it identifies what cannot be changed (noise factors) but then seeks to engineer robustness against these factors by using what can be changed (control factors), transforming yourself so that unchangeable circumstances do not produce negative effects.

The prayer offers a path of graceful surrender – "I cannot change this situation, so I will find peace with how it impacts me" – while the essay offers a path of internal reengineering – "I cannot change this situation, so I will change myself until it cannot disturb my peace." One is therapeutic, the other is transformative.

For example, if someone insults you – and talking does not help and walking away is not an option – the Serenity Prayer approach means "I feel hurt, but I accept that people sometimes say hurtful things and I will not waste energy being angry about it." You acknowledge the hurt but do not fight against it, so that the hurt does not turn into suffering.

The *Sthitaprajna* approach means "I hear the words, but they don't actually disturb my inner state because I am not dependent on others' opinions for my well-being." You have cultivated an internal attitude where the insult simply does not register as painful – like water rolling off a duck's back.

The two approaches may also be considered complementary in that if one approach fails the other may succeed.

Both approaches agree on "the courage to change the things I can" rejecting passive resignation. The Serenity Prayer does not advocate giving up, nor does the *Sthitaprajna* ideal suggest withdrawal from the world. Instead, both point toward active engagement – using our internal resources (wisdom, attitude, spiritual practices) to remain effective despite external turbulence.

Both approaches require the same diagnostic skill – recognizing what lies beyond our control. The "wisdom to know the difference" becomes a systematic assessment: Can this disturbance be meaningfully addressed externally, or do I need to build internal robustness against it? Or do I need to accept it? This transforms spiritual platitudes into an actionable decision-making framework – exactly what is needed for navigating life's inevitable challenges.

Summary

This essay presents a fascinating parallel between a robust product – that performs optimally regardless of noise, and the Hindu philosophical concept of *Sthitaprajna* – a person of steady wisdom who remains unperturbed by life's ups and downs. Through a dialogue between an engineer and a statistician, the essay illustrates the common principles by which both robust products and wise individuals achieve equanimity despite disturbances they cannot control.

The key insight lies in the mathematical concept of interaction between what we can control and what we cannot. We previously discussed this concept under practiced causality in the chapter "Cause and Effect." In product design, engineers seek factors that interact with noise to minimize their disruptive effects. Similarly, a *Sthitaprajna* uses internal factors like attitude, self-awareness, and discipline to neutralize the effects of life's inevitable disturbances.

A *Sthitaprajna Equation* is developed to show how happiness can remain constant despite increasing noise when one cultivates the right attitude. The essay emphasizes that this approach does not mean avoiding difficulties, or withdrawing from the world, but rather building internal robustness to skillfully manage disturbances.

This is not philosophy as abstraction – it is a systematic method for inner robustness. It begins with identifying which challenges demand external solutions, and which invite internal experimentation. Then, through iterative learning, one discovers the most effective physical and mental "design parameters" for neutralizing life's noise.

In essence: Hindu philosophy becomes a robust design manual – for the soul.

Note: To illustrate that the foundations of Hindu philosophy are rooted in practical experience, this essay first introduced the concept of Robust Product Design and then drew a parallel to the transformation of a human being into a Sthitaprajna. While Robust Product Design is a recent innovation – emerging only in the last few decades – the ideal of Sthitaprajna, and the disciplined methods to become one, have been elaborated in Hindu texts for millennia. This comparison underscores a striking insight: the principles of resilience and optimization, central to modern product design philosophy, were already embedded in the spiritual and psychological frameworks of Hindu thought thousands of years ago.

Chapter 10
Purposeful and Happy Life

Imagine a young doctor working long hours in an underfunded hospital. She chose medicine to serve others – a life of purpose. But years into her career, she feels burned out, disconnected, and quietly wonders: Is this what fulfillment feels like?

Now picture a successful entrepreneur who has built a thriving company. He has wealth, freedom, and recognition. Yet he wakes up each morning with a nagging emptiness: Is this all there is?

People often strive to achieve two fundamental goals in life: (a) lead a purposeful life of contributions to society, and (b) be happy. Conventional thinking says that we act only when we desire something, and we are happy only when those desires are fulfilled. This essay exposes the paradox at the heart of that logic: the quest to maximize purpose sabotages happiness and vice versa.

Based on secular reasoning and mathematical modeling, the essay shows that simultaneous maximization of happiness and purpose requires a fundamental shift in our thinking – how we relate to desire, action, and satisfaction. The analysis leads to two transformative strategies: desireless action – instinctive right action, and unattached action – where effort is wholehearted but emotionally free from outcomes. Together, these paths offer a practical framework for individuals to lead more purposeful and happy lives, and for society to achieve greater prosperity, happiness, and equality.

In doing so, the essay affirms the central insight of the Bhagavad Gita – Karma Yoga – not as religious doctrine, but as a timeless and universally applicable guide to personal and societal transformation.

Two Goals in Life

The desire to alleviate suffering and attain happiness is universal – understandably so, as the pursuit of happiness often is life's most natural goal. But should it be our *only* goal? The answer, on both ethical and practical grounds, is no.

First, a life oriented solely around personal happiness risks becoming self-centered, overlooking our broader responsibilities to others and the society that sustains us. Second, fixating on a singular goal can lead to unintended harm, especially when it crowds out other essential aims. A balanced approach – one that embraces multiple values – offers a more integrated and harmonious path through life.

Consider the Federal Reserve in the United States. Its role is not limited to curbing inflation. If that were its sole mandate, it could raise interest rates aggressively to tame price growth. But doing so might plunge the economy into a recession, fueling widespread unemployment. To prevent such imbalances, the Federal Reserve holds a dual mandate: price stability *and* maximum employment. One goal checks and complements the other.

Or take the design of a medical stent. If engineers were told only to ensure the device does not collapse under pressure, they might create a rigid steel plug – perfectly strong but obstructive to blood flow. In this case, durability alone is not enough. Ensuring uninterrupted circulation is vital. The success of the device depends on the interplay between both objectives.

In life, too, personal happiness requires a counterpart: the aspiration to make the world better for others – be it family, community, nation, humanity, or planet. Such a goal acknowledges the unpayable debt we owe to the social and cultural ecosystems that shaped us. It reflects a moral intuition that our lives are enriched not just by what we receive, but by what we give forward.

For this discussion, we shall distill life's aims into two core objectives, to be pursued within an ethical and lawful framework:

- A purposeful life, rooted in meaningful contributions to society.
- A happy life, grounded in personal joy and inner contentment.

Are these goals entirely independent? Not quite. They are partially interwoven. Purpose can often bring happiness, just as happiness can fuel purposeful living. And yet, the relationship is not guaranteed. History records many selfless changemakers – visionaries, artists, scientists – who gave immensely to society but suffered deeply on a personal level, some ending their lives in despair. At the same time, many find joy in private, tranquil lives without visibly shaping the wider world. Thus, while the goals intersect, they are not synonymous, and progress toward one may not assure success in the other.

Let us now consider how these two goals – happiness and purpose – can be clearly defined and meaningfully measured.

Measuring Happiness and Contribution

It is a common experience that we feel happy when our desires are fulfilled and unhappy when they are not. However, measuring happiness is far from straightforward. Not all fulfilled desires yield the same degree of happiness, and more importantly, happiness does not depend merely on the number of desires satisfied – it also hinges on the total number of desires entertained. For instance, if a person has only one desire and it is fulfilled, his happiness could be considered complete. Conversely, if someone has four desires and only two are satisfied, they may feel only partially happy. Thus, happiness is better understood as a proportion: the ratio of satisfied desires to total desires.

Guided by this intuition, we adopt a working definition inspired by Swami Chinmayananda, a renowned teacher of Hindu philosophy, as presented in *Self-Unfoldment* (Chinmaya Publications, 1994). For the sake of simplicity, it assumes that each satisfied desire contributes equally to happiness. Accordingly, happiness – scaled between 0 and 1 – can be expressed mathematically as:

$$\text{Happiness} = \frac{\text{Number of desires satisfied}}{\text{Number of desires entertained}} = \text{Probability of satisfying a desire} \dots (10.1)$$

Swami Chinmayananda also observes that fulfilling a desire often gives rise to new ones, perpetuating a cycle that rarely reaches lasting contentment. Despite dramatic advances in science, technology, and material comfort, average societal happiness may not have increased significantly over the decades – because what has changed is not necessarily our ability to fulfill desires, but the nature and scale of the desires themselves. Wanting a large house instead of a hut does not, by itself, raise the likelihood of fulfillment.

Turning to the concept of contribution, we observe that our positive impact on society stems from successful, purposeful actions. And practically speaking, purposeful action originates from desire – the desire to help, to create, to solve, to serve. Thus, our contribution to society is a function of the socially beneficial desires we hold and manage to fulfill. While desires may differ in their societal value, we simplify the model by assuming each fulfilled, contribution-worthy desire generates the same value k. We can then define an individual's total contribution as:

$$\text{Contribution} = k \, (\text{Number of desires satisfied}) \quad \dots \dots \dots \dots (10.2)$$

Together, these formulations create a useful framework for thinking about happiness and contribution – not only as subjective experiences or ethical ideals, but as measurable and interrelated outcomes shaped by desire, action, and probability.

Traditional Approach

In pursuing the twin goals of a happy and purposeful life, we naturally rely on our practical experiences. It is commonly observed that we feel happy when our desires are fulfilled and unhappy when they are not. Similarly, our ability to contribute meaningfully to society often stems from desires – specifically, the desire to help, serve, or create. If such desires are satisfied, we contribute; if they remain unfulfilled, our contributions diminish. Thus, the traditional approach to life centers on the fulfillment of desires.

Consider a simple example. Imagine a young man graduating from college who has two prominent desires over the next six months: securing his dream job, and marrying his ideal partner. Here, the number of desires is two. Suppose each has a 50% chance of being fulfilled, and that the outcomes are independent. Then, we can expect three scenarios: (1) Both desires are fulfilled. This occurs with 25% probability, yielding 100% happiness. (2) Only one desire is fulfilled. This occurs with 50% probability, resulting in 50% happiness. (3) Neither desire is fulfilled. This occurs with 25% probability, producing 0% happiness – or intense dissatisfaction.

If an individual bases his happiness on the percent of desires satisfied, then the individual will experience one of these three happiness levels with the probabilities outlined above. While the example assumes independence between desires for simplicity, real-world desires often exhibit correlations – success in one area can influence outcomes in another.

This framework mirrors how many people navigate life: alternating between fulfillment and disappointment, with happiness rising and falling in tandem with external outcomes. At a collective level, this model helps explain how happiness is distributed across society. For instance, among thousands of students with similar aspirations, we can expect a 25-50-25 percent split across the three outcomes. This results in an average happiness of 50% in society, with individual experiences ranging from total dissatisfaction to complete fulfillment.

What emerges is a statistical portrait of happiness – one shaped not only by effort or merit but by the inherent uncertainty in outcomes. Despite having the same goals and working with similar diligence, individuals experience varying levels of happiness due to the probabilistic nature of desire fulfillment. This inherent variability underscores the limits of the traditional, desire-based approach to achieving happiness and purpose.

Societal Consequences

If we continue to operate under traditional thinking – where desires drive action, and fulfilled actions drive happiness and contribution, as in Equations (10.1) and (10.2) – what will this imply for the distribution of happiness and contribution across society? What will the personal and policy implications be? To explore this, we construct a mathematical model that simulates this scenario and illustrates its broader implications. *While this section draws on basic probability and statistics, readers less inclined toward mathematics may skip to the next section – Strategies for Purposeful and Happy Life.*

Mathematical Model
The model envisions a typical timeframe – a year – during which each individual harbors several meaningful desires. The nature and quantity of these desires vary from person to person, as does the probability of success, influenced by individual capability and context. Each successful desire not only generates happiness for the individual but also contributes, in some way, to society. The sum of these successful outcomes determines the overall distribution of happiness and societal impact within the population.

We now build a simple model for illustrative purposes only adopting two key simplifying assumptions. First, we treat all individuals as identical in terms of the number of desires they entertain, the probability of fulfilling each desire, the happiness derived from fulfillment, and the societal contribution each fulfillment represents. While clearly an oversimplification, this assumption allows us to establish a conceptual baseline. Second, we assume that beyond a certain threshold, increasing the number of desires (n) leads to a proportional decrease in the probability of success (p), such that their product (np) remains constant. This reflects the intuitive constraint that time, energy, and attention are finite: stretching us too thin reduces the likelihood of success in any one pursuit.

It is important to note that while we assume uniformity across individuals for the purposes of simplification (more complex and realistic models can be built), this framework can be more appropriately applied within relatively homogeneous subgroups – such as socioeconomic strata – where variation is less pronounced. Even within these groups, happiness and contribution levels will vary, since they depend on how many desires are fulfilled. Thus, the model naturally reflects real-world disparities.

To formalize the discussion, we introduce the following notation:

- n is the number of desires entertained per person over a fixed period.
- x is the number of desires satisfied out of n entertained.
- p is the probability of satisfying any single desire, estimated as x/n.
- k is the societal contribution of each satisfied desire.
- CV is coefficient of variation, defined as standard deviation divided by mean.

The CV metric is essential for comparing variability of outcomes with different scales. While standard deviation gives an absolute measure of spread, CV reveals relative variability – particularly relevant in comparing variability of happiness and contribution.

If x, the number of fulfilled desires per individual, is assumed to follow a Poisson distribution with parameter np, then the conclusions are as follows (proof not shown):

- Average contribution per person = k(np)......................(10.3) since each person, on average, satisfies np desires, and each desire contributes k units to society.
- Person-to-person variability of contribution = CV (contribution) = $1/\sqrt{np}$ ....(10.4) reflecting greater variability when np is small.
- Average happiness per person = p...............................(10.5) since happiness is defined as the probability of satisfying a desire, estimated as the proportion of fulfilled desires.
- Person-to-person variability of happiness = CV (happiness) = $1/\sqrt{np}$ ....(10.6) reflecting greater variability when np is small. Note that under our assumptions, the CV of contribution and CV of happiness are equal.

These results underscore an important insight: even when individuals start with similar goals and constraints, the stochastic nature of desire fulfillment introduces variability in both happiness and contribution. This variability reflects a structural limitation of the traditional desire-driven approach – a limitation that becomes especially pronounced when scaled to the level of society.

Example

To illustrate the model in action, let us consider a scenario in which everyone entertains eight meaningful desires (n = 8) over a given period. Suppose the probability of satisfying each desire is 0.5, or 50%. In this case, p = 0.5 and np = 4. The resulting metrics, from Equations (10.3) to (10.6) are straightforward:
- The average contribution per person becomes 4k
- The coefficient of variation (CV) of contribution is 0.5
- The average happiness per person is 0.5
- The CV of happiness is also 0.5

These figures offer a statistical snapshot of how happiness and contribution might be distributed across society. For example, the average contribution per person is 4k, and the standard deviation of contribution is $(0.5)(4k) = 2k$. Similarly, average happiness is 0.5 and standard deviation of happiness is $(0.5)(0.5) = 0.25$, suggesting considerable variability in individual outcomes. Though all individuals begin with the same number of desires and identical probabilities of success, their experiences of happiness and contribution can differ markedly – reflecting life's inherently probabilistic nature.

Implications

If the above model were a good approximation of reality, several important implications would emerge – insights that could guide both personal choices and public policy.

First, there is an inherent trade-off between happiness and contribution. If individuals seek to maximize happiness, they will limit themselves to very few, easily achievable desires that increase the probability of satisfying a desire (p). However, this would lead to limited societal contribution. Conversely, increasing the number of desires may raise contribution but can lower p, thereby reducing happiness. Maximizing both happiness and contribution simultaneously is not feasible under conventional thinking.

Second, the model reveals a contribution plateau. As the number of desires increases, np and average contribution initially rises but only up to a point. When np becomes constant, contribution stagnates. Further increases in the number of desires (n) merely dilutes attention and effort, causing the probability of success (p) to fall, and happiness to decline. This illustrates the risk of spreading oneself too thin.

Third, the CVs for both happiness and contribution are equal, and inversely proportional to $\sqrt{(np)}$. As n increases, np initially increases and CV decreases – suggesting reduced inequality in outcomes. When np becomes constant, CV stabilizes.

These findings carry policy relevance. If the model is assumed to broadly reflect human behavior, it suggests that social and economic policies should be designed in tandem. Rather than simply generating more opportunities (increasing n), it may be more effective to focus on raising the probability of success (p) for the opportunities people already have. Enhancing p might mean investing in education, support systems, or reducing systemic barriers. Further, policies should guard against the artificial inflation of desires, which may reduce happiness without increasing fulfillment.

Limitations

The above model has severe limitations because it oversimplifies a complex reality. The assumption that np remains constant is a key driver of the contribution plateau, and treating all individuals as identical disregards the rich variation in desires, capabilities, and contexts. In practice, disparities in outcomes would be even greater than the model predicts. While more elaborate models could account for such diversity, *the current version suffices to illustrate how such modeling could be done, and its potential personal and policy implications assessed.*

We are now ready to address the pivotal question: what kind of inner and outer transformations are required to enable the simultaneous maximization of happiness and contribution – goals that are incompatible within the traditional desire-driven framework?

Strategies for Purposeful and Happy Life

Let us revisit Equation (10.1).

$$\text{Happiness} = \frac{Number\ of\ desires\ satisfied}{Number\ of\ desires\ entertained} = p \quad \ldots\ldots\ldots\ldots(10.1)$$

This formulation equates happiness to the probability (p) of satisfying a given desire. It implies two strategies for increasing happiness: either increase the numerator – by fulfilling more desires – or reduce the denominator – by entertaining fewer desires. From a purely mathematical standpoint, the most effective way to maximize happiness is to reduce the number of desires to zero. With no desires entertained, there are none left unfulfilled, resulting in a conceptual happiness of 100%. However, the nature of this happiness is no longer about fulfillment but reflects an inner equanimity – a serene, undisturbed state of mind.

Yet, zero desires also mean no desires can be fulfilled. According to Equation (10.2), this would result in zero contribution to society.

Contribution = k (number of desires satisfied) .... (10.2)

This creates a profound dilemma.

Dilemma
If we adhere to the conventional belief that happiness depends upon the probability of satisfying a desire, and contribution depends upon the number of desires satisfied, then happiness and contribution cannot be maximized at the same time. Maximizing happiness requires minimizing desires, which reduces contribution. Conversely, maximizing contribution demands entertaining many desires, which inevitably results in many unfulfilled ones – leading to frustration and diminished happiness. We are left with difficult questions: Should we minimize desires to pursue happiness? Maximize desires to pursue contribution? Aim for a compromise that balances the two? Or is there a more transformative approach?

A change in thinking is required to enable simultaneous maximization of both happiness and contribution. Specifically, we must revise either Equation (10.1) by redefining the source of happiness, or Equation (10.2) by redefining the source of action. We explore two strategies to do so: acting without first needing a desire (desireless action) and acting without attachment to the results (unattached action).

Desireless action represents a more advanced state – where right action flows naturally, without the ego-desire mechanism even engaging. The second strategy is more practical for most people – you may still want things and work toward them, but your inner state is not held hostage by whether you get them. Here, happiness arises not from success, but from action free from dependency on results.

These two strategies offer a radical departure from traditional thinking – one that may hold the key to living a life that is both deeply joyful and meaningfully impactful.

Desireless Action

One strategy for simultaneously achieving maximum happiness and societal contribution is to retain Equation (10.1), which defines happiness as the ratio of satisfied desires to total desires, and revise Equation (10.2), which currently links contribution to fulfilled desires. According to Equation (10.1), the most effective route to perfect happiness is to reduce the number of desires to zero. With no desires entertained, there is no gap between desire and fulfillment – resulting in a conceptual 100% happiness characterized by equanimity and inner peace. However, using the traditional Equation (10.2), zero desires mean zero satisfied desires implying zero contribution. To escape this paradox, we must redefine Equation (10.2) so that *action – and thus contribution – need not originate from desire*. This leads us to the idea of desireless action.

Unlike selfish or even selfless actions, which are still rooted in desire – either for personal gain or for others' welfare – desireless actions arise spontaneously, unprompted by calculated thought. They emerge from a transformed state of being. Like instinctively catching a falling child or rendering help without deliberation, these actions flow from an individual whose nature has aligned so deeply with truth and compassion that the right response arises naturally. There is no ego seeking reward, no fear of failure, no deliberation – only clear, unburdened action.

To further clarify, we can think of human actions across three broad categories: selfish, selfless, and desireless. Selfish actions are driven by personal gain; selfless actions elevate societal welfare over self-interest but still originate from higher-order desires. When such desires go unfulfilled, they may still result in disappointment. Desireless actions, however, transcend both. They do not originate from will or want but from a sense of innate harmony with deep integration of thought, feeling, and action. Such individuals act not *out of* desire or duty but *as* the embodiment of rightness. Their contribution flows effortlessly, like the fragrance of a flower – neither strained nor strategic.

By moving toward this ideal of desireless action – closely approximated by truly selfless action – one can achieve both enduring happiness and maximum contribution. If the mathematical model in the previous section is directionally accurate, such a way of being would also reduce disparities across individuals, fostering a more equitable and integrated society.

Unattached Action

An alternative strategy is to preserve Equation (10.2) – which rightly links societal contribution to the number of satisfied desires – while redefining Equation (10.1), so that happiness no longer depends on desire satisfaction. Unlike the "desireless action" approach that seeks to eliminate desires altogether, this path – unattached action – accepts the presence of desires and encourages active engagement with them. *The key, however, is not to anchor happiness in the outcomes.*

In this revised framework, happiness becomes a function of mindset rather than achievement. Any definition of happiness that (a) does not rely on desire fulfillment, and (b) is easily achievable, would suffice. Among many such formulations, those that *also* promote societal contribution are the most noble and transformational. Two notable variations stand out under this approach.

The first is faith-based happiness – a deep and abiding trust in something larger than oneself. For instance, if one wholeheartedly believes in a benevolent and personal God, then happiness can be defined as: *Happiness = Faith in God*. This is the essence of *Bhakti Yoga*, where happiness arises from surrender and devotion. Even in the face of adversity, faith sustains peace. Bad outcomes are seen not as personal failures but as divinely orchestrated events, freeing the individual from anguish and allowing them to continue acting without emotional volatility.

However, while this method supports inner stability and happiness, it does not *require* outward action. A person can remain blissfully content in solitude or inaction, which may be unsatisfying to those who view contribution as essential to a meaningful life.

The second is unattached action, the traditional *Karma Yoga* path. Here, happiness is derived not from achieving results, but from acting purposefully and ethically, without emotional dependence on the outcome. The guiding equation becomes: *Happiness = Pursuit of desires to contribute*. In this model, happiness lies in the sincerity of effort, not the certainty of success. The journey becomes the destination. By aiming to fulfill as many contribution-oriented desires as possible – while remaining inwardly free – individuals optimize societal impact while safeguarding personal peace and happiness. Moreover, because success is no longer the yardstick of happiness, disparities between people naturally decline, fostering a more harmonious social fabric.

Both *Bhakti Yoga* and *Karma Yoga* can lead to deep fulfillment. But for those who seek not only inner peace but also external impact, *Karma Yoga* stands out. It cultivates happiness through engaged action and transforms service into joy. By uniting contribution with inner freedom, it offers a compelling framework for living a purposeful and happy life.

Practical Applications – *Karma Yoga*

Our secular, logical, and mathematical exploration has led us to two complementary strategies for simultaneously maximizing individual happiness and societal contribution: desireless action and unattached action. Remarkably, this conclusion aligns with one of Hinduism's most profound teachings – *Karma Yoga*, as presented in the *Bhagavad Gita*. This ancient text articulates *Karma Yoga* as the disciplined path of right action without attachment, a concept that resonates directly with the principles derived from our analysis.

Bhagavad Gita places particular emphasis on unattached action – performing one's duties wholeheartedly, yet without fixation on the outcome. To a lesser extent, it also gestures toward desireless action, a more advanced state where action arises naturally, free from both ego and craving. These two modes of being are respectively captured in verses 2.47 and 2.71 of the *Bhagavad Gita*:

You have control over your actions,
but not on their fruits.
Act for action's sake,
do not be attached to inaction.

Abandoning all desires,
acting without craving,
free from the sense of 'I' or 'mine'
he attains peace.

Karma Yoga, then, is not an abstract ideal but a deeply practical approach to life. It integrates the discipline of non-attachment with the intuitive spontaneity of desireless service. The journey begins with ethical restraint – avoiding actions that cause harm. From there, one progresses from selfish action (driven by personal gain) to selfless action (motivated by collective welfare), to unattached action, and to desireless action (spontaneous right action), where one acts not from compulsion or obligation, but from an inner alignment with truth. At this stage, action is not a function of desire, but a reflection of character – it flows as effortlessly as a river runs or a flower gives fragrance.

This progression affirms a foundational insight: happiness need not be contingent on success or failure, and contribution need not be predicated on desire. By freeing oneself from the emotional volatility of outcome-based identity, *Karma Yoga* empowers individuals to act with vigor, purpose, and serenity. Its dual emphasis on *engaged detachment* and *effortless action* gives it extraordinary transformative potential.

As this essay has shown, even within a secular and analytical framework, *Karma Yoga* emerges as a compelling paradigm. Its adoption at scale could yield profound societal benefits. Individuals would experience greater happiness and fulfillment, while their contributions to the collective would multiply. A society grounded in *Karma Yoga* would not only see a rise in measurable prosperity – reflected in metrics like Gross Domestic Product (GDP) – but also experience a deeper, more sustainable rise in collective well-being, akin to what one might call Gross Domestic Happiness (GDH). Moreover, disparities in happiness and contribution would diminish, fostering social cohesion, ethical harmony, and a more inclusive economic ethos – within the realistic boundaries of institutional policy.

Summary

This essay presents a secular and mathematical framework for addressing a timeless question at the heart of human existence: how can one simultaneously achieve personal happiness, and make meaningful contributions to society? It identifies a fundamental tension embedded in conventional thinking – namely, the belief that happiness depends upon the probability of satisfying a desire, and contribution depends upon the number of satisfied desires. This creates an inherent conflict: maximizing happiness calls for minimizing desires to increase the probability of satisfaction, while maximizing societal contribution demands the pursuit of many ambitions, thereby lowering the likelihood of fulfillment and diminishing contentment.

Within this framework, happiness is defined as the probability of satisfying a desire, and estimated as the ratio of fulfilled desires to total desires. Contribution is measured as the total number of fulfilled desires, each weighted by its societal value. Using simplifying assumptions, the essay demonstrates how mathematical modeling can reveal the systemic consequences of traditional desire-driven behavior and can inform the design of policies that aim to enhance both personal well-being and societal impact.

To resolve the inherent dilemma associated with conventional thinking, the analysis points to two complementary strategies that converge remarkably with the ancient Hindu philosophy of *Karma Yoga*. The first is desireless action – spontaneous, instinctive behavior not driven by conscious desire, but arising from a natural alignment with truth and goodness. Such action maximizes contribution while preserving happiness through equanimity, independent of external success. The second is unattached action, where individuals actively pursue meaningful desires but remain emotionally free from the outcomes. Fulfillment arises not from achievement, but from wholehearted engagement in the pursuit itself.

Both approaches, constituting *Karma Yoga,* represent a profound reorientation of how we typically relate to desire, action, and satisfaction. If widely embraced, *Karma Yoga* could enable the simultaneous flourishing of individuals and societies. It offers a model in which economic growth, as measured by Gross Domestic Product (GDP) and holistic well-being, reflected in what might be called Gross Domestic Happiness (GDH) rise together, while disparities in both happiness and contribution narrow. Most strikingly, this framework – arrived at through secular logic and mathematical reasoning – affirms the core teaching of the Bhagavad Gita: act righteously, without attachment to results. This convergence of ancient wisdom and modern analysis reveals *Karma Yoga* not only as a spiritual path, but as a universally applicable strategy for a productive, equitable, and fulfilled human life.

Glossary

1. *Adharma* - Actions, thoughts, or behaviors that go against moral and ethical principles. The opposite of *dharma*. Not doing one's duty. *Adharma* creates negative *karma* and moves one away from spiritual progress.
2. *Advaita* - "Nondual" or "not two." *Advaita Vedanta* teaches that the universe and universal consciousness (*Brahman*) are fundamentally identical, with apparent separation being an illusion created by *Maya*.
3. *Ananda* - Pure, unlimited bliss or joy that transcends ordinary happiness. One of the fundamental aspects of *Brahman* (along with *Sat* and *Cit*). This bliss arises naturally when one realizes one's true divine nature, and is not dependent on external circumstances for happiness.
4. *Ananta* - Literally "without end" – uncountably infinite. When applied to *Brahman*, it indicates infinite in space, time, and objects.
5. *Anantam* - Grammatical form of *Ananta*, meaning the same uncountably infinite, endless nature of ultimate reality.
6. *Asankhya* - Countably infinite, like the infinite sequence of natural numbers (1, 2, 3...). Philosophically and mathematically, this represents a smaller infinity compared to *Ananta*.
7. *Atman* - The true Self or essential nature of every individual. Pure, unchanging consciousness that makes all experiences

possible. In *Advaita Vedanta*, *Atman* is identical with *Brahman* – It is the *Brahman* within you. The realization of this identity is the goal of spiritual practice.

8. **Bhagavad Gita** - "The Song of God" – an 18-chapter, seven hundred verse philosophical dialogue between Prince Arjuna and Lord Krishna on the battlefield of *Kurukshetra*. This sacred text addresses fundamental questions about duty, action, devotion, knowledge, and the nature of reality, making it one of Hinduism's most influential works.

9. **Brahma Sutra** - Also known as *Vedanta Sutra*, this text by sage *Vyasa* systematically examines the nature of *Brahman* through concise aphorisms. Along with the *Upanishads* and *Bhagavad Gita*, it forms the "triple foundation" of *Vedantic* philosophy.

10. **Brahman** - The ultimate, absolute reality underlying all existence. Infinite in space, time, and objects. Described as *Sat-Cit-Ananda* (existence-consciousness-bliss). It transcends all qualities and descriptions while being the ground of all being.

11. **Brahmin** - One of the four *varnas* (social classes). In the traditional *varna* system, this is the social class dedicated to learning, teaching, and spiritual duties. *Brahmins* were responsible for preserving sacred knowledge, conducting rituals, and maintaining *dharmic* principles in society. Members include priests, teachers, thinkers. Originally based on spiritual qualities rather than birth.

12. **Causal Body** - Contains a unique proportion of the three *gunas* – *sattva, rajas, tamas* – that everyone has, and which manifests as tendencies. Also contains past *karma*. Causal body carries forward from life to life and determines one's fundamental disposition and destiny.

13. **Chatur Varna** - The "four *varnas*" or the original social classifications based on an individual's qualities and natural tendencies: *Brahmins* (knowledge-seekers), *Kshatriyas* (protectors), *Vaishyas* (producers/traders), and *Shudras* (service providers). Ideally based on merit and temperament, not heredity.

14. *Cit* - Pure consciousness itself – not consciousness "of" something, but consciousness as the fundamental nature of reality. One aspect of *Brahman*, representing the knowledge principle that illuminates all experience. Pronounced Chit.
15. *Dharma* - One's righteous duty or the natural law that maintains universal and social harmony. *Dharma* operates at multiple levels: universal principles (like truthfulness), social duties (based on one's role), and individual purpose (*svadharma*). Following *dharma* leads to spiritual growth and societal well-being.
16. *Dvaita* - "Dualistic" philosophy, particularly associated with sage *Madhvacharya,* and emphasizing devotion. However, in this book, we have used a modified version of *Sankhya Philosophy*, with its infinite *Purushas* replaced by a single infinite *Brahman,* as an example of dualistic framework without the consideration of personal God.
17. **Free Will** - The capacity to choose between alternatives. In Hindu philosophy, while past actions influence present circumstances, individuals retain the ability to choose their current actions, and shape their future.
18. *Gunas* - The three fundamental qualities or forces that constitute *Prakriti*: *Sattva* (harmony, purity, clarity), *Rajas* (activity, passion, restlessness), and *Tamas* (inertia, ignorance, dullness). In humans, these appear as psychological tendencies that influence behavior and spiritual development.
19. *Iti* - "Thus" or "so it is" – used to mean "in this manner" or used as a concluding statement.
20. *Jagat* - The universe as it appears to ordinary perception. In *Advaita Vedanta, jagat* is considered *mithya* (apparent reality) – real at the practical level but dependent upon and an appearance of *Brahman*.
21. *Jivanmukta* - A person who has attained liberation (*moksha*) while still living in a physical body. Such individuals have realized their true nature as *Brahman* and live in constant bliss, free from the suffering caused by identification with the limited self.

22. *Jiva* - The individual Soul or empirical self – *Atman* when limited by the three bodies (physical, subtle, causal) and identified with personal experiences, thoughts, and *karma*. The *Jiva* experiences pleasure and pain, birth and death.
23. *Jivatma* - Another term for *Jiva*, emphasizing the Soul aspect of individual existence.
24. *Jnanam* - Spiritual knowledge or wisdom that goes beyond intellectual understanding to direct realization of truth. This experiential knowledge transforms one's entire perspective and leads to liberation from ignorance and suffering.
25. *Karma* – Action. Also, the universal law of cause-and-effect governing actions and their consequences. Includes: *sanchita karma* (accumulated from past lives), *prarabdha karma* (destiny for this life), *agami karma* (being created now), and *systemic karma*. Understanding *karma* helps explain life circumstances, and the importance of right action.
26. *Krishna* - One of Hinduism's most beloved deities, considered an avatar of God Vishnu. As Arjuna's charioteer and teacher in the *Bhagavad Gita*, Krishna represents divine wisdom guiding the individual Soul through life's challenges toward spiritual realization.
27. *Kshatriya* - One of the four *varnas*. The warrior and ruler class, responsible for protection, governance, and upholding justice. *Kshatriyas* were expected to embody courage, leadership, and sacrifice for the greater good while maintaining ethical principles in their authority.
28. *Mahabharata* - The world's longest epic poem, narrating the complex story of two royal families (Pandavas and Kauravas) and their climactic war. Beyond the narrative, it contains profound philosophical teachings, including the *Bhagavad Gita*, and is considered a comprehensive guide to *dharmic* living.
29. *Mahavakyas* - Great sentences. *Tat Tvam Asi* – That Thou Art – is one example.

30. *Manana* - The process of deep reflection, contemplation, and analysis that follows initial learning (*shravana*). *Manana* involves questioning, reasoning, and internally digesting teachings until intellectual understanding becomes firm conviction.
31. *Maya* - The mysterious projecting and veiling power of *Brahman* that makes the One appear as many. *Maya* simultaneously veils *Brahman* and projects the appearance of a diverse universe.
32. *Mithya* - Apparent reality – neither absolutely real (like *Brahman*) nor completely unreal (like a fantasy). The world and individual experiences are *mithya*: they appear real but depend on *Brahman* for their existence, like waves depend on the ocean. As another example of dependent reality, the dream state is less real than the waking state, which is less real than the ultimate reality *Brahman*.
33. *Moksha* - Liberation from the cycle of birth, death, and rebirth. It is the ultimate goal of human life. *Moksha* involves realizing one's true identity as *Brahman*, and becoming free from all limitations, suffering, and the compulsion to be reborn.
34. *Namaste* - A respectful greeting meaning "I bow to You" or "the divine in me honors the divine in you." This salutation acknowledges the sacred essence present in every being, and promotes humility and spiritual recognition in daily interactions.
35. *Neti* - "Not this" – a method of inquiry (*neti, neti*) used to discern what *Brahman* is by systematically negating what it is not. By eliminating all finite attributes and experiences, one points toward the infinite, indefinable nature of ultimate reality.
36. *Nishkama* - Acting without attachment to results not for personal gain. *Nishkama karma*, taught in the *Bhagavad Gita*, involves performing one's duties with full effort while surrendering the outcomes to the divine, not basing our happiness only on successful results, because success is not entirely in our hands.
37. *Parinama* - Real transformation where the cause actually changes to become the effect, like milk becoming yogurt.

Contrasted with *vivarta* (apparent transformation), *parinama* is used in *Sankhya* philosophy to explain how *Prakriti* evolves into the material universe.

38. **Prakriti** - Primordial matter or the material cause of the universe, composed of three *gunas*. Under the influence of consciousness (*Purusha*), *Prakriti* evolves into all forms of matter and energy, creating the diversity of the physical world.

39. **Pramana** - Valid means of knowledge recognized in Indian philosophy, such as perception, inference, testimony. Last word on a subject.

40. **Rajas** - One of the three *gunas* – the *guna* of activity, passion, and restlessness. In psychological terms, *rajas* manifests as ambition, desire, emotional turbulence, and constant doing.

41. **Rajasic** - Characterized by *rajas guna* – active, passionate, ambitious, but also restless and potentially agitated. A *rajasic* person is driven, goal-oriented, but may struggle with peace and contentment.

42. **Rigveda** - The oldest of the four *Vedas*, containing hymns to various deities. These ancient Sanskrit verses form the foundation of Hindu ritual and philosophy, celebrating the divine forces of nature, and establishing key spiritual concepts.

43. **Sankhya** - One of the six classical schools of Hindu philosophy, founded by sage Kapila. *Sankhya* is dualistic, positing two fundamental realities: *Purusha* (pure consciousness, infinite in number) and *Prakriti* (infinite primordial matter). It provides a detailed analysis of creation and the path to liberation through discrimination between these two realities. In this book we have considered the infinite *Purushas* as a single *Brahman*.

44. **Sat** - Pure existence or being-ness – the fundamental "is-ness" that underlies all objects and experiences. As one aspect of *Brahman*, *Sat* indicates that absolute reality is self-existent and the source of all existence.

45. **SatCitAnanda** - The compound word describing *Brahman's* essential nature as Existence-Consciousness-Bliss. These three are

not separate ingredients of *Brahman*, but three aspects of the one indivisible absolute homogeneous reality that is the ground of all being.

46. ***Sattva*** - One of the three *gunas* – the *guna* of harmony, purity, and clarity. *Sattvic* qualities include peace, wisdom, compassion, and spiritual insight. Even though it is the highest of the three *gunas*, *sattva* must ultimately be transcended to realize *Brahman*.

47. ***Sattvic*** - Characterized by *sattva guna* – pure, harmonious, wise, and spiritually inclined. A *sattvic* person naturally gravitates toward truth, service, and spiritual practices, experiencing greater peace and clarity.

48. ***Satyam*** - Truth in both its absolute sense (*Brahman*) and relative sense (honesty, factual accuracy). *Satyam* is both a fundamental principle to be realized, and a virtue to be practiced in daily life.

49. ***Sharira*** - Body, referring to the three levels of embodiment: gross physical body (*sthula sharira*), subtle body of mind and intellect (*sukshma sharira*), and causal body of *gunas* and *karma* (*karana sharira*).

50. ***Shloka*** - Verse or stanza in Sanskrit poetry.

51. ***Shravana*** - Attentive listening with an open, receptive mind prepared through proper qualifications. The first step in the classical path of knowledge.

52. ***Shudra*** - One of the four *varnas*. In the *varna* system, those dedicated to service and labor in support of society. Originally based on temperament and aptitude for practical work rather than birth, with service being considered a noble path when performed with dedication.

53. **Soul** - Same as *Jiva*. Causal and subtle bodies enlivened by *Atman*. The individual conditioned consciousness that experiences life through multiple incarnations, carrying forward *gunas* and *karma*, and evolving toward ultimate realization of its true nature as *Atman*.

54. *Sthitaprajna* - A person of steady wisdom who has achieved equanimity and Self-realization. Such individuals remain balanced in success and failure, pleasure and pain, maintaining clarity and peace regardless of external circumstances.
55. **Subtle Body** - Humans have gross, subtle, and causal bodies. Our gross sense organs – ear, skin, eyes, tongue, and nose interact with the corresponding five subtle bodies resulting in five types of sensations. Our mind and intellect are subtle bodies. The subtle body in its totality survives physical death and carries forward into subsequent incarnations.
56. *Tamas* - One of the three *gunas* – the *guna* of inertia, ignorance, and dullness. While *tamas* can provide necessary rest and stability, *tamasic* tendencies include laziness, confusion, destructive behavior, and spiritual darkness.
57. *Tamasic* - Dominated by *tamas guna* – characterized by lethargy, ignorance, procrastination, and destructive tendencies. *Tamasic* individuals struggle with motivation, clarity, and positive action.
58. *Tat Tvam Asi* - "That Thou Art" – arguably the most important statement of *Vedanta*. This profound teaching directly points to the identity between the individual Self (*Tvam*) and absolute reality (*Tat*), dissolving the apparent separation between individual consciousness and universal consciousness.
59. *Upanishad* - "Sitting near" (a teacher) – learning the final, philosophical sections of the *Vedas* containing the highest spiritual teachings. The major *Upanishads* explore the nature of *Brahman*, *Atman*, and the path to liberation through dialogue, story, and direct instruction.
60. *Vaishya* - One of the four *varnas*. The merchant and agricultural class in the *varna* system, responsible for economic production, trade, and wealth creation. *Vaishyas* support society through business acumen, resource management, and generating prosperity for the community.

61. **Varna** - The original fourfold classification system based on each individual's qualities, abilities, and spiritual temperament rather than birth. The four *varnas* are: *Brahmin, Kshatriya, Vaisha,* and *Shudra.* Each *varna* represents a different approach to serving society, and pursuing spiritual growth according to one's natural disposition.

62. **Vasana** - Subtle mental impressions or tendencies created by past actions and experiences in this and previous lives. *Vasanas* influence current thoughts, desires, and behavior, creating patterns that shape personality and destiny until altered through the application of free will.

63. **Veda** - "Knowledge" – the four ancient sacred texts (*Rig, Yajur, Sama,* and *Atharva Vedas*) that form the foundation of Hindu wisdom. Each *Veda* contains hymns, rituals, meditation techniques, and philosophical teachings covering all aspects of life and spiritual development.

64. **Vedanta** - "End of the *Vedas*" – both the concluding philosophical portions of the *Vedas* (*Upanishads*) and the systematic philosophy based on them. *Vedanta* explores the nature of reality and liberation, representing the pinnacle of *Vedic* wisdom. *Vedanta* is one of the six systems of Hindu philosophy, the others being *Sankhya, Yoga, Nyaya, Vaisheshika,* and *Mimamsa*.

65. **Vivarta** - Producing an effect without a change in cause, like seeing a snake in a rope. This concept explains how *Brahman* appears as the universe while remaining unchanged – the world is a *vivarta* (apparent transformation) of *Brahman*.

66. **Yoga** - "Union" – the various paths and practices that lead to the realization of one's true nature, and unity with *Brahman*. The four main *Yogas* are *Bhakti* (devotion), *Raja* (meditation/physical practices), *Jnana* (knowledge), and *Karma* (selfless action), each suited to different temperaments and approaches to spiritual growth.

www.ingramcontent.com/pod-product-compliance
Lightning Source LLC
Chambersburg PA
CBHW071659090426
42738CB00009B/1595